DISCARD
BETHANY
COLLEGE
LIBRARY

IMPURE SCIENCE

IMPURE SCIENCE

Fraud, Compromise and Political Influence in Scientific Research

Robert Bell

John Wiley & Sons, Inc.

New York • Chichester • Brisbane • Toronto • Singapore

Published by John Wiley & Sons, Inc.

Library of Congress Cataloging-in-Publication Data

Bell, Robert, 1942–
Impure science : fraud, compromise, and political influence in scientific research / Robert Bell.
p. cm.
Includes bibliographical references.
ISBN 0-471-52913-3 (cloth : alk. paper) : $22.95 (est.)
1. Fraud in science—United States. 2. Research—United States—Moral and ethical aspects. I. Title.
Q175.37.B45 1992
507.2073—dc20 91-43148

Printed in the United States of America

10 9 8 7 6 5 4 3 2 1

Printed and bound by Courier Companies, Inc.

To Isabelle

Acknowledgments

A number of persons have given me encouragement, advice, and help on this book. I would especially like to thank some personal friends and colleagues who were tremendously generous with their suggestions, concerns, time, and effort: Paul Pavel, whose kindness, wisdom, and entire life put me in a mind to write this book; my agent and friend Jonathan Dolger, who insisted that I pursue this book after I had given up on it; my editor at Wiley, Roger Scholl; my friends Tom Fuchs, Hy Sardy, Larry Bell, Barry Pascal, Robert Higashi, Marek Kanter, Jacques and Suzy Plat, Shuli Buttler, Peter Chenko, and Gerry Haggerty.

Of course, I'm responsible for what I've said in the book, but without the help of all of the above, and many others, there would either be no book or much less of one.

CONTENTS

INTRODUCTION

Impure Science documents how certain members of the scientific community have encouraged a trend toward political influence, compromise, and fraud in scientific research. Although the overwhelming majority of scientists have not been direct participants, they have remained silent onlookers or have looked the other way.

This trend in science is a matter of great concern since it affects, among other things, the medicines we take, the bridges we drive on, the buildings we live in, and the weapons that are supposed to defend us. Such malfeasance and compromise subverts our scientific and technological base, thereby weakening the competitiveness of the economy in which we earn our living.

The book is based on original sources, including documented evidence from court cases, government investigations, testimony given under oath, Freedom of Information Act requests, and personal interviews. I have analyzed a number of cases in different scientific fields, to demonstrate that these examples are part of a pattern. My primary purpose is to show that the American scientific community is as "pure" and unbiased as the political machinery that dispenses its patronage and its funding.

HOW MUCH FRAUD IN SCIENCE?

Daniel Koshland, Editor of *Science,* thinks fraud in science is rare. He stated in an editorial that "99.9999%" of scientific papers "are accurate and truthful."[1] William Broad and Nicholas Wade, the authors of a widely discussed book on science fraud, take the opposite view. "We would expect that for every case of major fraud that comes to light, a hundred or so go undetected. For each major fraud, perhaps a thousand minor fakeries are perpetrated."[2] Dr. Robert Sprague, who paid dearly after he exposed a science fraud affecting mentally retarded children, states, "There is very little evidence on the frequency of fraud in general. We simply don't know."[3] The evidence suggests, however, that the fraud cases that come to light are but a small fraction of the whole. A study at a major university showed that of the 32% of researchers who suspected that a colleague had falsified data, 54% did not report it or try to verify their suspicions.[4] In fact, the same year (1987) that Koshland's editorial appeared in *Science,* the journal also published an article based on purloined data, according to a 1989 report by the National Institutes of Health. "The NIH report describes a situation that is truly unfortunate for all concerned," said Koshland.[5]

FOLLOWING THE MONEY

Scientists cannot conduct scientific inquiries without money. In reality, obtaining federal grant money and being subject to federal regulation are characteristics that help identify members of the scientific community.

A recent report by the Office of Technology Assessment (OTA) puts this into context, stating, "There will always be more opportunities than can be funded, more researchers competing than can be sustained, and more institutions seeking to expand than the prime sponsor—the Federal Government—can fund."[6] Actually, the pressure on the funding system and on those trying to obtain money is increasing. In 1989, only 31% of proposals submitted to the National Science Foundation (NSF) were funded. Ten years earlier, the figure was 38%.[7] At a conference on scientific fraud, Frank Solomon, a professor of biology from the Massachusetts

Institute of Technology, stated: "Good young people and good older people are going without funding, and it creates an environment which is amenable to all sorts of corner-cutting."[8] He went on to explain that the stakes today are much higher than when he was an undergraduate: "For example, no one ever thought that you could get rich doing biology, and now I'm the only biology professor without a Saab. . . . The whole tempo has changed. It is possible to get wealthy doing biology. It is possible to become quite well-known doing biology. And our students learn those lessons quickly."

Today's laboratories often have a well-known scientist at the top, who can win grants from the funding agencies and who supervises many younger associates. There is often an emphasis on producing relatively large numbers of scientific papers with many coauthors (some of whom may not really contribute to the works). The laboratory head and certain associates may even get an essentially honorific listing on the papers as a sort of academic payola.

The pressure is for greater quantity, rather than for reduced output that would improve quality. As the OTA report states, "[S]ince most overhead is brought into the university by a small number of research professors (at Stanford, 5 percent of the faculty bring in over one-half of the indirect cost dollars), proposals to reduce research output are not looked on with favor by many university administrations."[9]

Because of this situation, university officials, who are scrambling for funds, and top officials at government funding agencies, who are fighting to protect their budgets, do not encourage the unfavorable attention that whistle-blowers may bring to the university or the agency. Not only can bad publicity give an edge to rival institutions competing for money, but the university exposed by a whistle-blower might actually have to forfeit money it has already received or been promised. "Under current policies, a finding of misconduct can result in the institution's probable loss of a grant; for that reason, institutions have a financial incentive to conclude that their faculty are not guilty of misconduct," wrote Congressmen John Conyers and Ted Weiss to congressional colleague John Dingell in a letter dated June 17, 1991.

Since money is indispensable to conduct science, part of the plan of this book will be to "follow the money." The path that money follows to and from institutions shows the exact roles played by funding agencies, university officials, and private corporations.

WHAT THE EVIDENCE SHOWS

Three mechanisms are supposed to prevent behavior that can lead to fraud and compromise in scientific research. The first is the peer review system, by which committees of specialists in a field determine the merit of the individual proposals submitted for funding. The second is the referee system, by which scientific papers are sent out for review to qualified reviewers who recommend whether the papers should be published, changed, or rejected. The third mechanism is the replication of experiments. In theory, other scientists attempt the same experiment; if they cannot replicate the results, the claimed result is dismissed. Replication of experiments, however, rarely takes place; there is almost no money available for it, or in it. Why should peer reviewers allocate some of the limited funds to duplicate what has already been done? Also, a scientist who simply replicates other researchers' results would soon be labeled a hack and wouldn't get further funding. Instead, rival scientists try to establish their reputations by extending the results in some way.

Replication attempts do occur *within* laboratories, because it may be relatively cheap for a skilled researcher to repeat a fully equipped experimental procedure. Postdoctoral fellow Margot O'Toole's inability to replicate some experiments that laboratory head Dr. Tereza Imanishi-Kari claimed to have conducted ultimately led to the much publicized Baltimore scandal discussed in Chapter 4.

The peer review systems, while sound in theory, are prone to abuse. The secrecy that surrounds the process creates a potential for conflict of interest from reviewers competing for the same funds; researchers whose proposals are rejected also feel pressure not to question the system.

Even journal peer review is vulnerable to abuse. In several documented cases, journal editors have refused to send out for peer review articles that criticize other scientists' work or indicate possible fraud.

Grant peer review, journal peer review, and experiment replication are supposed to make science self-correcting. But *Impure Science* illustrates that preeminent people involved in science are repeatedly defeating these mechanisms. If science is at all self-correcting in the United States, it is despite the efforts of some of these powerful individuals, not because of their efforts.

CONFLICTS OF INTEREST, COVERUPS, AND ABUSES OF TRUST

Several patterns often recur in investigations of scientific research. One has to do with conflict of interest—when the scientist who is supposedly making an objective judgment stands to benefit or lose by that decision. Conflict of interest appears at all levels of government and private scientific organizations; it is a fundamental fact in Pentagon-funded research. The Pentagon's senior scientific advisory group, the Defense Science Board, for example, has at most a handful of practicing scientists; it consists overwhelmingly of senior executives from major Pentagon contractors.

Another current pattern in science is *concurrency,* which is defined as the practice of performing fundamental research and development while simultaneously mass producing the item being researched. Because military contractors make so much more money in the production of a weapon system than in its research and development, weapons go into mass production long before these preliminary stages have been completed. As a consequence of this compromise, recently "developed" U.S. weapons often fail to work effectively. Asked by an aide to Congressman Dingell at a closed door hearing to name a single Gulf War weapon that came on-line during the arms buildup of the 1980s that actually worked "without quality problems," a top Pentagon official replied, "I will be very honest. I do not have one on the tip of my tongue." The same pattern of concurrency, which a congressional report characterized as an "earn as you learn" philosophy, occurs in medical, pharmaceutical, and other scientific research.

Perhaps nowhere is there greater abuse of the public's trust than in the "super" science projects directly funded by Congress. Such projects are "political" science by definition, since Congress is constitutionally designed to respond to political pressure. Too often, these projects far exceed budget and fail to work as planned. Among the more publicized scientific fiascos discussed in this book are the Space Shuttle and the Hubble Space Telescope. A project that has not received as much publicity—the nuclear "pumped" X-Ray Laser, which was the original centerpiece of Star Wars—may be the most shocking case of all.

One of the most extraordinary aspects of scientific funding is the treatment received by scientists who attempt to expose fraud in science. The whistle-blower frequently becomes the victim, and the subsequent "investigation" of fraud by the university or government funding agency becomes an investigation of the informant and a cover-up of the reported action.

The abuses in the name of science documented herein are surprising. Even more startling is the response to such abuses by the science agencies and the scientific community. At one hearing, Congressman Dingell summed up a common thread of his own investigations, ". . . the scientific community, which has apparently been treated as a sacred cow for rather long, acts as if the government is making it a victim of some kind of persecution if it is questioned about accountability for Federal funds. . . ."[10] It is my hope that *Impure Science* will help to banish the secrecy that surrounds science funding and will encourage reexamination of the conduct of science in today's world.

CHAPTER 1

Shortchanged: The Breakdown of Peer Review at the National Science Foundation

This is the first of two chapters about the role of politics and political influence in the funding of basic science research by the National Science Foundation (NSF). With a 1990 budget of $1.58 billion, the Foundation is only the fifth largest of the federal science funding organizations, but its influence on science is greater than its size would seem to indicate. The NSF awards the majority of grants in a number of fields in the natural and social sciences, from anthropology to zoology to engineering. In many of these fields, it is the only federal agency to which scientists can turn for funding for basic research.

This chapter examines how the NSF funds individual scientists involved in traditional, relatively small-scale projects. (The next chapter will look at questionable NSF practices in awarding grants for large projects.)

I have concentrated on a specific case in this chapter to document the kinds of flaws that exist in the NSF's grant selection process. In this instance, the Foundation got caught; the case discloses conflicts of interest on the part of the peer reviewers, shows how decision making is influenced by gossip and innuendo, and illustrates the devastating effects such gossip has had on one scientist's career.

NSF officials have argued that the case is exceptional. The Foundation's General Counsel, Charles Herz, has written, "Those involved in preparing or reviewing this response know of no other problem NSF has faced quite like that one."[1]

Although some details in *Kalb v. National Science Foundation* are surely exceptional, two facts related to the case raise serious doubts about the uniqueness of some NSF policies and practices. First, apparent conflicts of interest were repeatedly ignored, over a period of years, by higher Foundation officials.

Second, there is no way to substantiate the words of the NSF officials. For more than a decade—in apparent violation of the Federal Privacy Act—the NSF maintained a secret and illegal filing system, designed deliberately to hide and withhold critical information from grant applicants. Because of the violation of the Privacy Act, the Foundation's Inspector General referred the Foundation to the Justice Department for possible criminal prosecution, although the Justice Department declined to prosecute.

This case documents how the peer review system can, and to varying degrees does, break down in actual practice. "I feel that such abuses are inherent in the system and that by definition they are pervasive, to one degree or another," said Jon Kalb, the man whose career was damaged by the NSF's actions, in written testimony to a congressional task force.[2] A year later, the NSF admitted to a string of abuses in the case, publicly apologized to Kalb, and paid him $20,000 to cover his legal costs.

PEER REVIEW

Peer review often determines how federal agencies decide who gets the money to do scientific research and how much researchers get. The definition of peer review differs at different agencies, and sometimes within the same agency. The basic concept is that experts in a particular field review the research proposals of other researchers under the assurance that the reviewers' identities will remain anonymous. At the National Institutes of Health (NIH) the person applying for funding knows the identity of the judges but is never informed how any particular judge rated the proposal. This is the way most university promotion and tenure decisions are made as well. At the NSF, the cloak of secrecy surrounding the peer review system is complete; the applicant is never allowed to know who the judges are, let alone how each one responded. Without such secrecy, the NSF claims, reviewers would be afraid to offer frank opinions. As with letters of recommendation for job applicants, positive proposal reviews are unlikely to ricochet at the reviewer. But negative reviews might provoke an angry response, even legal action—or so the NSF claims.

Despite such justification, not everyone agrees with blanket anonymity in peer review. One congressman described peer review, as currently practiced, as "an 'old boy's' system where program managers rely on trusted friends in the academic community to review their proposals. These friends recommend their friends as reviewers. . . . It is an incestuous 'buddy system' that frequently stifles new ideas and scientific breakthroughs."[3] C. W. McCutchen, an NIH scientist who gave the Kroc Memorial Lecture at the Massachusetts Institute of Technology in 1990, described another problem with peer review when he said, in a telephone interview on May 10, 1990, "Your fate is decided by your competitors." Although peer review at the NSF has damaged his career, Jon Kalb says, "There's nothing wrong with the concept of peer review—it's essential; the problem is *abuses* of peer review."

Erich Bloch, former Director of the NSF, said about the peer review system: "Since its inception, the National Science Foundation has relied on the judgment of qualified, experienced scientists and engineers to select the most promising high quality research for support. This system . . . has been found to be remarkably effective and flexible, adapting to the changing needs of science and engineering research, and incorporating improvements with respect to openness, accountability, equity, and impact on the research environment."[4]

What was the nature of the peer review system used in the *Kalb* case? The NSF describes it as follows:

"Proposals are assigned to a program officer who oversees external reviews, evaluates reviewers' comments, and makes a recommendation to award or decline it, taking into account other considerations—such as the relationship of the work to the field as a whole and to other pending proposals, and the program's purpose and budget.

". . . the proposal is sent to several people . . . identified by the program officer as knowledgeable on the topic. The reviewer receives standard instructions and forms and responds directly to the program officer. Where panels are used, they meet . . . to weigh a group of proposals, taking into account the prior mail reviews. . . . All reviews are routinely sent to the proposer, but without attribution to individual reviewers. Where panels are employed, summaries of their deliberations are also provided. . . ."[5]

Jon Kalb describes the procedure a bit differently: "Reviewers of proposals are anonymous, and advisory panel meetings that evaluate proposals are closed meetings, so the influence of possible vested interests on the research proposal is unknown to the grant applicant. . . . [N]o official minutes of a standard nature are in practice required of these meetings, and the 'panel summary' may or may not reflect the actual ideas and recommendations of the panel. Instead, the NSF Program Director ultimately makes the final decision on proposals, subject to the approval of his or her superiors."[6]

THE BACKGROUND TO KALB V. THE NATIONAL SCIENCE FOUNDATION

From 1973 to 1978 Jon Kalb, who was in his mid-30s, researched prehistory sites in Ethiopia for new evidence of early humans. During these years, he made a major impact on the field. *Science,* the journal of the American Association for the Advancement of Science, described him as "a pioneer investigator" of human fossils in Ethiopia.[7] The distinguished French researcher Professor Maurice Taieb, who is now the Director of Geological Research for the Centre National de la Recherche Scientifique (CNRS; the French equivalent to the NSF), at Luminy near Marseille, has said "Kalb discovered the Bodo Man."

Although Kalb has made major archeological discoveries and published a number of scientific articles on paleontology in important journals such as *Nature,* he was, in fact, formally trained in neither field. A graduate student in geology at Johns Hopkins University, he went to Ethiopia in early 1971 to conduct geological mapping with the Ethiopian Geological Survey. By agreement with the Survey, he also investigated mineral prospects for a small mining exploration company. His geological work brought him into contact with Maurice Taieb, who was then doing fieldwork as a graduate student of the University of Paris. After collaborating in field explorations in late 1971, the two agreed to form a multidisciplinary research team, which they eventually called the International Afar Research Expedition (IARE). The famed anthropologist L. S. B. Leakey lent his prestige to the expedition by writing Kalb an endorsement letter for funding. The IARE later became famous for the discovery of the sensational "Lucy" skeleton.

Taieb and Kalb drew other researchers into their research plans, including Yves Coppens, of the Museé de l'Homme of Paris, another French graduate student, Raymonde Bonnefille, and Professor Karl Butzer, of the University of Chicago. Butzer did not participate in the expedition, but he and Bonnefille recommended to Taieb a University of Chicago graduate student in anthropology, Donald Carl Johanson.

In early 1972, following more field explorations by Kalb and joint surveys by Kalb, Taieb, Coppens, and Johanson, the four signed a protocol at the Ethiopian Antiquities Administration (EAA) officially forming the IARE. By mutual agreement, Taieb was named head of the expedition. At about the same time the EAA gave Taieb a "concession," an exclusive franchise to explore a 30,000 square kilometer area of northeastern Ethiopia in a desert region known as the Afar Depression, which includes the now famous Hadar and Middle Awash areas. These areas yielded the Lucy fossils and Bodo Man, respectively. Because of the concession, no one could explore or work any sites in the area without Taieb's expressed permission.

Johanson, with the assistance of Butzer and Kalb, applied to the NSF for money for the expedition's fieldwork. Kalb applied to a private foundation in Connecticut, the Fund for Overseas Research Grants and Education (FORGE), for the continuation and expansion of a small grant FORGE had already given him. Taieb and Coppens applied to the CNRS. All of them received some research money; Johanson's NSF grant even included modest funds for Kalb.

Unlike the other three, who returned to their respective institutions at the end of each fieldwork season, Kalb was a legal resident of Ethiopia where he and his wife raised their family from 1971 until mid-1978. He eventually rented a big house and compound in Addis Ababa to use as offices and as a staging center for the IARE.

Kalb's permanent residency in Ethiopia was one reason the other researchers were attracted to him. "Kalb was important because he was living in Ethiopia. He offered to provide a car, a house to stay," recalled Bonnefille, who is still Taieb's colleague, but now at Luminy, near Marseille. If Kalb started out with the group in 1972 as a convenience to young French and American researchers, by 1975 he was viewed as a full-fledged scientific rival.

Prehistory-related studies in East Africa are notoriously cut throat. Johanson and Kalb had grave differences about documentation of the fossil hominid finds made at Hadar by Johanson in late 1973, and by the way Johanson announced the finds to the press. Kalb and Johanson sent contradictory letters on the matter to the NSF. Kalb, in a July 3, 1974, letter to Mary Green, Assistant Director, Programs for Anthropology, at the NSF, wrote, "During a visit to the U.S. early this year, I sought a closed audience at the Wenner-Gren Conference in New York city with responsible peers of Johanson's profession to discuss the circumstances surrounding his late 1973 hominid discoveries: the fact that he withheld confirmation of the hominid finds from members of the Expedition . . . and that he allowed no neutral party to view the finds in the field, and that he blatantly broke a signed agreement by issuing a premature announcement of the finds to the press without informing other members of the expedition beforehand." Kalb elaborated on his claims concerning Johanson's fossil finds in a later report to the NSF, "It was not until the end of the field season, in mid-December [1973]—six weeks after the initial fossils had been found—that we learned that hominid-fossils had in fact been discovered at Hadar, a confirmation that we read about in the newspapers."[8]

Johanson gave his views on some of Kalb's charges in an undated 1974 letter to Dr. Iwao Ishino, of the NSF, in which he asserted that Taieb had left him, Johanson, in complete authority over the expedition when Taieb had temporarily gone back to France. Johanson argued that he was obliged to go along with the desires of Ethiopian officials in first announcing the discoveries in Ethiopia since they wouldn't let him take the fossils out of the country without the announcement. He also maintained that his arrangement with Taieb included that as soon as he got to Paris, a press conference with Coppens and Taieb would be held. His letter, cosigned by Taieb, did not refer to all of Kalb's charges. Although Kalb would continue (and still continues) to insist on the truth of all of his charges, apparently Johanson did not consider a number of them worthy of comment.

Part of Kalb and Johanson's dispute had to do with money. Johanson had a 24-month NSF grant for $44,400. Kalb's share of that grant was $3,750.[9] Johanson's letter to the NSF, characterized above, also argued that due to NSF funding cuts, he had had to cut Kalb's funding as well, reducing it to the period of the field season rather than the full year. Johanson said that Kalb "insists" on continuing to receive "full support" although the total grant had been cut by 60%.

Kalb admitted to the NSF that this was partially true. He claimed, however, that Johanson withheld details of this cut in budget money from himself, as well as Taieb, until a few weeks before the expedition assembled in Addis Ababa, even though Johanson had revised the budget himself five months previously.[10]

In his July 3, 1974, letter to Mary Green, of the NSF, Kalb stated a viewpoint that would come to foreshadow subsequent events: "It is my definite opinion that Johanson has sought to influence Ethiopian officials and even Afar native assistants to the detriment of others, for reasons of self-aggrandizement. This could affect my security in a country where foreigners already have a difficult existence, and possibly even the physical well-being of my field party who work alone in the Afar those months of the year when the expedition is not here. . . ." In mid-1974, Kalb renewed fieldwork in the Afar Depression accompanied by several Ethiopian and American students.

In a June 1, 1975, letter from Johanson to Maurice Taieb, Johanson stated that he had spoken to a researcher and colleague of Kalb from Southern Methodist University, Professor Fred Wendorf, who had told him that Kalb was about to submit an article for publication labeling key features of the Hadar research region. Such labeling by Kalb would be "very dangerous," Johanson indicated. He urged Taieb to publish "very quickly" an alternative article in France accomplishing much the same thing. This would be followed, Johanson proposed, by a second article after Taieb arrived for an expected visit to Cleveland. Johanson would have this article quickly published by the Cleveland Museum. With Taieb getting priority on the geographical labeling, he wrote, "Kalb has no chance."

In his undated 1974 letter to the NSF, described earlier, Johanson concluded that he and Maurice Taieb were working together in an attempt to find a disposition to the difficulties posed by Kalb.[11] Taieb confirmed his cooperation with Johanson in an

August 23, 1974, letter to Johanson in which Taieb wrote that he was determined to get the permit from the Ethiopian government for an expedition from which Kalb would be eliminated. Kalb felt that Johanson intentionally tried to drive a wedge between Taieb and himself. But in an interview, Taieb, who expressed great personal regard for Kalb, said that he thought Kalb was too competitive to remain on the expedition.

Because Kalb was a partner in the IARE agreement with the government, Johanson's dispute with Kalb meant that neither Johanson nor Taieb could get permission from the Ethiopian authorities to go into the field. In late 1974, Johanson requested that the NSF present a letter of support to the Ethiopian authorities. The NSF sent a telegram on September 13, 1974, effectively giving Johanson blanket support. As a result, the Ethiopian government gave Johanson, Taieb, and others permission to go into the field without Kalb. Shortly thereafter, Kalb resigned from the expedition that he had helped found.

Johanson reaffirmed his stance on Kalb in a letter he wrote to the NSF on October 8, six weeks later. He said that at least at that time he thought further research involving Kalb simply couldn't be conducted because more than once Kalb had tried to "block" the fieldwork and had also tried to "prevent" Johanson and others from pursuing their investigations.

Apparently, the NSF officials were satisfied with the outcome, which had caused Kalb to withdraw from the expedition. (In an earlier, September 17, internal NSF memo, the then Program Director Iwao Ishino had written, "This is the latest in a series of 'disturbances' that Mr. Kalb seems to be involved in." On September 19 he wrote to General Counsel, "My inclination [is] to discount" Kalb's side of the dispute.) Ishino then wrote to Johanson on November 19, 1974, "Many thanks for your letter of October 28, 1974, in which you describe Mr. Jon Kalb's present status with reference to the International Afar Research Expedition. I am relieved to know that the . . . [Ethiopian authorities have] reached a decision and that the expedition has received permission to carry out its fieldwork."

The NSF did not know, however, that Kalb was at that moment in the process of outflanking the IARE. Kalb's actions came as a surprise to many. Johanson had written Taieb several months earlier, in a letter dated June 11, 1974, that he found it peculiar that he and his colleagues hadn't heard anything concerning Kalb.

Johanson speculated that Kalb might be distressed with his excolleagues and perhaps was at a loss for words.

Kalb, however, knew exactly what to say, and especially to whom to say it—the Ethiopian government. He formed a new research organization, called the Rift Valley Research Mission in Ethiopia (RVRME). In early 1975, the Ethiopian government carved out of the concession they had previously given to Taieb a territory exclusively for the RVRME. It turned out to be one of the best territories in the entire region. Kalb stated: "I appealed to the Ministry [of Culture] to give my new team permission to study a central portion of the IARE permit area. The area I requested was a virtually unexplored region in the central Middle Awash Valley that was one-seventh the size of Taieb's total permit area; also it . . . had never been visited by the IARE since the expedition had been established. I bolstered my appeal with a report I submitted to the Ministry in January, 1975, that was a summary of my own explorations and mapping for the IARE over a three year period." Kalb also asked for permission to explore three smaller areas in the event that the Middle Awash area was denied. The Ethiopian officials gave the RVRME an exclusive study area permit on all of his requests. "He has a concession," an NSF official later wrote in an internal memo, "The Gov't gives them out as one would mining concessions. To keep it he must, in effect, start mining. . . ."[12] Kalb now had the franchise for an area that Taieb had previously held. If anyone else wanted to work there, they now had to work with Kalb's team.

Kalb wrote of the explorations he and his RVRME colleagues made throughout 1975 and 1976: "We made outstanding discoveries, including a fossil skull of an archaic hominid and perhaps the largest single Paleolithic archeological site in the world—35 square kilometers of continuous artifact occurrences. We also found fossil sites much older than the Lucy site, by several million years. There were other scientific teams that were foaming at the mouth to get into our study area." Part of Kalb's argument to the Ethiopian authorities was that he would do field studies in the area, whereas his rivals might never get around to it. "I tried to argue that the work just at Hadar [the site which by then had yielded the sensational "Lucy" skeleton] could keep the IARE busy for 30 years," wrote Kalb.[13]

Why did Ethiopia give Kalb and the RVRME the concession? Kalb offered something to the desperately impoverished country that no other researchers then provided. Kalb's new organization was a permanent base for research in Ethiopia, rather than an expedition that floated in and out of the country. It had year-round offices in Addis Ababa and a permanent Ethiopian staff, and conducted year-round fieldwork. Kalb recalled: "We were the first such prehistory organization in Ethiopia. . . . We established the first paleontology research laboratory in the country, at Addis Ababa University. There were 25 members of this organization, half of them were Ethiopians. . . . We had a model training program for Ethiopian students. Seven students from Addis Ababa University wrote their BS theses based on research with the RVRME, which in turn helped five of these students earn Fulbright scholarships for graduate studies in the U.S. Four of these students then received Masters degrees and three of them PhDs. These Ethiopians are the real pioneers of *Ethiopian* prehistory studies." To finance the research and overall operation of the RVRME, Kalb and his new associates raised funds, entirely in small grants, from 14 different foundations and research institutions in the United States and Europe. The funds were collectively pooled to pay for field expenses and the training of Ethiopian students.

Kalb argued that both his former colleagues in the IARE and the NSF had objectives that were vastly different from his own: "Their idea was to take the money and run, so to speak, with regard to working in Ethiopia. Since I was a resident of the country, I felt that working with Ethiopians and Ethiopian institutions was of extreme importance. Johanson went on to fame and fortune through his Ethiopian discoveries and raised substantial funds to establish his own institute in California, the Institute for Human Origins in Berkeley. That's terrific for Johanson and terrific for California, but it did nothing for Ethiopia." An NSF official later stated that NSF's "role is to enhance science in the United States, not to enhance science in a host country." Having said that, he did acknowledge that "in a science of this nature, the two are very much intertwined."[14]

KALB AND THE ETHIOPIAN DERG

The Emperor Haile Selassie, "King of the Kings, Elect of God, Conquering Lion of the Tribe of Judah," was overthrown in a

1974 military coup led by officers who ultimately revealed themselves to be Marxists. The revolution had followed a period of mass famine, which Kalb himself witnessed daily while doing fieldwork in the hardest hit parts of the country. The new government, with its avowed leftist platform and radical land reform, abolished all private ownership of land. This triggered widespread violence. "[M]any regions of rural Ethiopia became the scene of protracted social conflict as old and new groups fought for power and land. Thousands were killed," wrote Marina Ottaway, a university professor in Ethiopia and her husband, David, a *Washington Post* correspondent stationed in the country.[15] The government itself was unstable, with reports of murderous gunfights actually breaking out between government ministers at cabinet meetings. The situation was so bad that the central government initially kept itself secret. Only an inner circle knew the identity of the real leaders of the government. The popular name of the government reflected this unusual fact. It was known as the *Derg*, Amharic for "shadow."

Kalb was sympathetic to the early idealism that the government leaders espoused. "Please don't portray me as an adversary of what it was that the Ethiopian reform government was trying to accomplish. . . . When that revolution took place in Ethiopia, they overthrew 2,000 years of feudalism, and they really tried to set up a populist government. So this was the climate at the time I established RVRME."

KALB AND THE NSF

At the end of 1976, Kalb's research team applied for NSF money in three separate proposals. Kalb was not the principal investigator in any of them. He and other researchers from the RVRME were simply listed as part of the proposals, although Kalb coordinated the overall research project. The principal RVRME investigators were all from major American universities and included prestigious American researchers. Two of the proposals were sent to the Anthropology Program at the NSF. The first proposal involved archeological investigations in the now famous Middle Awash Valley, to be headed by Professor Fred Wendorf, of Southern Methodist University (SMU). A second proposal, on paleoanthropology studies, was to be headed by Professors Glenn

Conroy and Clifford Jolly, of New York University. The third proposal, on paleontological and geological studies, was headed by Professor Bryan Patterson, of Harvard University. It was submitted to a different NSF program, the Systemation Biology Program, in early 1977. Kalb's role in all three proposals, however, was later conceded in an internal NSF document written by John Yellen, the NSF Program Director for anthropology from summer 1977 to the present. Yellen wrote that the three "proposals compliment each other. All three projects will take place in the same area and Kalb will be the de facto director of all of them."[16]

Kalb's presence may have been enough to put the kibosh on the proposals. They were all turned down.

A CONFLICT OF INTERESTS?

The Foundation's Program Director for the Anthropology Program in 1976, Dr. Nancie Gonzalez, was also the chairperson of the advisory panel that evaluated anthropology proposals and was responsible for selecting outside experts to evaluate the SMU and NYU grant applications. The NSF professes to seek reviews from a broad spectrum of individuals. Yet, the NSF's own documents show that three of the five outside reviewers chosen to review the Wendorf proposal were from the University of California at Berkeley. This connection would later prove to be a major conflict of interest. Although Donald Johanson, Kalb's major rival, did not teach at Berkeley, documents will show that he was very close to those who did, ultimately establishing his own research institute in Berkeley. In fact, one of the Berkeley reviewers, according to a lawsuit Kalb later filed, "had very close ties to [Kalb's] competitors, and also had his own continuing grant application pending with the Anthropology Program."[17]

"There is a big difference between someone who has a professional interest in a proposal and someone who has a direct, blatant and overlapping interest in it," said Kalb. The Foundation itself recognizes that this can be a problem. NSF guidelines, which every reviewer must now sign, include the following clause: "If your designation gives you access to information not generally available to the public, you must not use that information for your personal benefit or make it available for the personal benefit of any individual or organization."[18] Kalb, however, has pointed out this restriction, which was not even in force at the time of the peer reviews that directly concerned him, is a limp-handed restriction at best. "Even today, NSF conflict of interest guidelines say almost nothing about the most pervasive type of conflict of interest—conflict of *research* interest."

What happens if someone violates the letter or spirit of the signed conflict of interest statement? "Well, I don't know that we ever take anybody to court," said Jim McCullough, the Foundation's Director of Program Evaluation Staff, in an interview in his office on May 11, 1990. Kalb, in turn, has asked, "What relief is given to a grant applicant, after a reviewer with a vested interest in a proposal misuses proprietary information in it or intentionally tries to kill it with a bad review in order that the reviewer enhances his or her ends?"

One of the reviewers of the SMU proposal was Professor Glynn Isaac of the University of California at Berkeley, who was a revered figure in the field, having been Richard Leakey's senior archeologist. There is some evidence for Kalb's belief that Isaac had close ties to competitors of Kalb. In an April 7, 1975, letter to Maurice Taieb, Donald Johanson, then still with the Cleveland Museum of Natural History, wrote that he had met with Glynn Isaac as well as Richard Leakey in New York the previous weekend and had talked with another Berkeley anthropologist, Clark Howell. Johanson indicated that he found out the Ethiopian authorities had given Kalb permission to research the Afar region. According to the letter, Desmond Clark, who was another Berkeley anthropologist, had stated that Kalb would be conducting his research with SMU Professor Fred Wendorf. Johanson said that he didn't see any way "we" could "stop" Kalb and Wendorf at this point, and that "everyone here" is quite "unhappy" about Kalb's getting his authorization to work in the Afar. He added that he had no idea as to what action they could take that could affect the situation. Clark Howell had previously headed the paleoanthropology program at the University of Chicago, where Johanson earned his PhD. Desmond Clark had his own expedition working in southern Ethiopia.

In a legal complaint, Kalb later alleged that Isaac wrote NSF Program Director Nancie Gonzalez to warn her about a rumor concerning Kalb that the panel, as well as higher NSF officials, should take up. In a later telephone conversation, Isaac elaborated on the rumor.[19] The peer review panel at this time was considering the SMU and NYU proposals. The Foundation acknowledged the truth of Kalb's allegations in a formal reply to Kalb's complaint:

"In a letter dated February 23, 1977, Glynn Isaac wrote the then Director of NSF's Anthropology Program . . . a letter in which he noted that there were rumors concerning Kalb's status."[20] The rumors concerned Kalb's alleged CIA involvement. Isaac's letter stated, "Kalb stayed on in Ethiopia after his formal employment in the Geological Survey had come to an end, and this gave rise to various speculations and rumors. As far as I am concerned the whole thing was a tissue of hearsay—but the rumors were of a kind, which must be of concern to the international community of anthropologists. Perhaps, at some point, Kalb should be invited to repudiate the rumors." Although Professor Isaac thought the rumors were "a tissue of hearsay," he elaborated on them and suggested that higher NSF officials consider them. In its answer to Kalb's complaint, the NSF admitted that, in a later conversation, "Isaac told Gonzalez that the rumors were that Mr. Kalb was associated with the Central Intelligence Agency (CIA)."[21] It took the NSF almost a decade to concede that "the rumor has no basis in fact. No documentation or any other type of evidence has ever been produced to support the CIA rumor."[22]

Isaac's review, however, was clouded not only by his ties to Johanson but also by what Kalb felt was a conflict of interest. Isaac had a continuing NSF grant, a relatively large award through the NSF's peer review process, that provisionally ran for three years. The first year, 1976, he was awarded $80,800. The letter awarding him the money stated, "Contingent on the availability of funds and the scientific progress of the project, it is our intention to continue support at approximately the following level: 2nd increment $91,300."[23] He had to apply for this money but did not need to go through peer review a second time. The research had to justify a second funding, but a scientist as distinguished as Professor Isaac would presumably be able to show sufficient progress to achieve that requirement. But there also had to be enough money in the NSF pot to win a second increment. As Kalb has pointed out, "We were requesting something like $250,000 and that would have taken a very sizable chunk of money out of that program. Also word had gotten around that we had made monumental discoveries in the Middle Awash Valley." Kalb felt that because Isaac had a continuing award pending that was also for archeological research in East Africa, he should not have been asked to review the RVRME-related proposals. Is it a mere coincidence that Isaac sent his letter mentioning the CIA allegations to the NSF's

Program Director for Anthropology on February 23, 1977? On March 11, 1977, Isaac sent the NSF a "request for support" for a grant for $87,135. Whether or not there was a connection, the potential for conflict of interest should have been avoided.

In his February 23 letter, Isaac recommended support for the NYU and SMU proposals that he had reviewed. He wrote, "I consider both good proposals. Wendorf's is excellent. . . . I feel that both projects should be supported." But he also pointedly referred, in his recommendation, to the "tissue of hearsay." This, combined with his follow-up phone conversation with the NSF elaborating on the rumors about Kalb, raised questions, at least in Kalb's mind about Isaac's intentions. "His *formal* review of the proposal gave it an 'excellent,' rating" said Kalb, "but then he also submitted the 'informal' secret review."

Because of Isaac's high professional standing, the NSF Program Officer could not ignore the allegations.

Is it possible for a reviewer to succumb to temptation where a conflict of interest arises? Obviously yes, but whether this is what happened in this case is unclear. Kalb has pointed out, "Whatever he did, whether it was a response to those facts or not, I don't want to make a judgment on it. He's the one who took the CIA thing to NSF, with a letter and follow-up conversation. I'm not about to make a judgment on Isaac's motives." As for Isaac's ties to Johanson or other researchers working in Ethiopia, Kalb said, "I feel he was being used by those with a greater vested interest than his own." Of course, even the appearance of a conflict of interest could have been avoided if the NSF had chosen a different reviewer.

Kalb interviewed Isaac on October 21, 1982, and made the following detailed notes of their conversation: "His motivation [in relaying the rumor] was to have [them] cleared up; I said Gonzalez's information was very specific, beyond just casual rumors. Covert funding was not casual. He said I did not have institutional affiliation and a clearly salaried position. Many people knew this. This was part of rumor, it came packaged with CIA issue—how else could I be living in Ethiopia over the long term, was the question he reported; [Isaac] cannot identify individual sources. He had heard the rumor from all working in Ethiopia; if he can do anything let him know."

In the NSF's decision-making process, the Program Officer not only sends out proposals for review to outside reviewers but, for many NSF programs, also gets the advice of a panel that meets to discuss the proposals. Sometimes the panel votes after discussing the comments of the outside reviewers; at other times they do not take a formal vote but discuss the proposal until reaching a consensus. The outcome of their vote or discussion usually determines the disposition of the proposal. The Program Officer, however, is not at all bound by the views of the outside reviewers or even the votes or consensus of the panelists.

Did the panels reviewing the proposals learn of the rumor? After rejecting the SMU and NYU proposals in mid-1976, NSF officials formally denied any evidence of this. But when Kalb sued the Foundation, the NSF's General Counsel finally acknowledged in a court settlement that the panel *did* learn of the allegations. "The NSF conveyed the CIA allegations to peer review panels which were responsible for reviewing the proposals," states the court settlement.[24] The document goes on to say that the Foundation "did not intend that its communication of the CIA rumor to the peer review panels would be considered an endorsement of the rumor, but some panel members have indicated that their views on the proposals were affected by the rumor."[25] The settlement acknowledged that the CIA rumor was taken under advisement by panel members reviewing the proposals, although "this item is not reflected in summaries of the panel reviews which were prepared by NSF."[26] According to Kalb, "The panel records were thus 'sanitized.'"

Imagine the thoughts of the panel members reviewing Kalb's proposals. The panel receives notification from the NSF of a shocking but unsubstantiated piece of gossip. The source of the gossip is a famous scientist in the field. Is it likely the panel members would ignore such allegations?

In addition, NSF then leaked its knowledge and concerns about the rumor to the academic community. Fred Wendorf, the man who would have headed the SMU project, heard the rumor as well. "He was actually told by Gonzalez of the significance of the rumor on the decision," said Kalb, "but NSF repeatedly denied this formally." The SMU professor wrote to Kalb on May 25, 1977, "In regard to your queries on how this information was conducted to me: it was done personally while I was in Washington and in considerable detail." On May 4, 1977, Wendorf had written to Gonzalez, "It smacks too much of McCarthyism, where if someone calls you something often enough, no matter how often you deny it, there remains a cloud of doubt. This is unfair. . . ."

Kalb wrote the NSF in 1978 requesting documents about the CIA allegations under the Freedom of Information Act (FOIA). At first, the NSF ignored his requests, but in the 1980s when the full duplicity of the Foundation's actions became apparent, he began bombarding the NSF with FOIA requests in an attempt to reconstruct what actually had occurred. In the end, Kalb managed to find out who the panelists were. Then he interviewed them and was able to reconstruct a fairly detailed account of the panel's deliberations. In a June 7, 1983, letter, Kalb broke down the discussions into 13 points. In its efforts to respond to the letter and an accompanying FOIA request, the Foundation's General Counsel's Office assigned one of its attorneys to call former panelists to confirm or refute Kalb's 13-point summary. This staffer wrote down a brief notation summarizing the results of the investigation for each of Kalb's points. Kalb then obtained this record through yet another FOIA request.[27] The result, to Kalb's astonishment, was a confirmation from the Foundation of his reconstruction of the panel meeting that had discussed the SMU and NYU proposals reviewed by Glynn Isaac. The results are reproduced here verbatim, followed in each case by the verbatim notes taken by the Foundation staffer who verified their accuracy:

1. Gonzalez began the Wendorf [SMU] panel meeting by reading the Isaac letter. *NSF notation:* "Correct."
2. Gonzalez elaborated on the Isaac letter by introducing the subject of the CIA allegations. *NSF notation:* "Correct."
3. These allegations were central to all following discussions concerning the proposal. *NSF notation:* "Probably true—Just before lunch. Discussion was very short and primarily on large issue (CIA)."
4. Gonzalez as explicit that NSF's support of anyone engaged in CIA activities would blacken the name of NSF and would discredit the professional community. *NSF notation:* "True—Certain."
5. The scientific merit of the Wendorf proposal was not discussed in any substance. *NSF notation:* "True—Nancie said she wanted to check with State Dept. and then get back to it."
6. Wendorf's involvement in Egypt was mentioned only marginally to the CIA issue, and that in any case it was felt that

Wendorf was professional enough to handle both projects. *NSF notation:* "Wendorf's Egypt work was discussed and was not considered a problem."

7. No panel member was given the opportunity to make a presentation on the Wendorf proposal. *NSF's notation:* "Can't remember one way or the other."
8. No vote was made by the panel on the scientific merits of the proposal. *NSF notation:* "At that time they tended to categorize in one or more category and try to reach a consensus. Voting was uncommon. Recollection was Nancy said to put it off until she checked with State."
9. The political situation in Ethiopia was not discussed in any depth in regard to either the Wendorf or Conroy [NYU] proposals, nor was it considered a central issue. *NSF notation:* "Can't recall."
10. The Wendorf panel meeting ended with the Wendorf proposal being "tabled" by Gonzalez, and her telling the panel that she would make inquiries with the State Department and/or the CIA concerning my status. *NSF notation:* "Yes."
11. The Conroy [NYU] proposal was linked with the Wendorf [SMU] proposal. *NSF notation:* "Yes."
12. The panel did not write a "panel summary"of either proposal. *NSF notation:* "Can't recall, but if they did it would have been pro forma."
13. Neither proposal was discussed by the panel in any future panel meetings in the Anthropology Program. *NSF notation:* "Yes. Think Conroy [NYU] discussion was on merit but much discussion on link with Wendorf. Was put aside because they were tied together. He thinks they were tabled for this reason but isn't sure. Can't remember what was recommended."

In an additional note, the NSF attorney also recorded: "There was a cocktail party at Nancie Gonzalez's one evening and the CIA was discussed among a few of them. He isn't sure if [the Assistant Program Director] was there. She was at a dinner one night. He thinks that [name deleted] participated in this cocktail discussion."

The conclusions of the NSF lawyer corroborated an earlier, 1982, internal NSF report that has never been publicly released. In 1982 Kalb appealed his case to a fellow Texan, Vice President

George Bush. Bush's office formally requested an account of the case from the Director of the NSF. The request worked its way down the bureaucratic ladder until it stopped on the desk of Dr. Charles Redman, who in 1982 was the newly appointed Associate Director of the Anthropology Program at the NSF. He interviewed several of the panel members, who gave him their recollections of what had happened. Redman wrote of one panelist, ". . . he said that Nancie had started the consideration of the Wendorf proposal by reading to the panel Isaac's letter bringing up the accusations and then saying that we can't fund someone like that."

Regarding another panelist, Redman wrote, "She recalled [the panel meeting] by asking, 'Was that the one where Fred [Wendorf] had a functionary over there in the Company?' I said yes, that was the one, but it was only an allegation. She said yes, there was a lot of discussion at the panel and that Nancie and one of the panelists seemed to know quite a bit about the CIA business and others had heard of it, but she was in the dark until then. . . . [She] believes that it was turned down largely because of the CIA implication. No written documentation was presented about the allegation. I asked if there was State Department pressure not to let people go to Ethiopia and she said there was not. . . . [She] said the proposal got an 'A' rating, but was tabled (and she said tabled in a way she thought it would not come up again)."

Redman also interviewed the former Program Director, Nancie Gonzalez, who was then with the University of Maryland. "She said she could not remember in detail and said she had thought of taping panel meetings at the time. She did recall that she made lots of phone inquiries and could not get any official government denials. She anguished over the situation and talked to many people, and the two major questions brought up to her were Kalb's source of funds and his ability to move freely around the country and seemingly close relations with all sorts of Ethiopians. She was not sure whether allegations had been brought to her attention in a letter or not, but if the letter had been addressed to her personally, she would not be surprised if it did not enter the official file. Some more direct statements from Gonzalez, 'If he was with the CIA it would be a terrible thing for the Foundation,' 'We took great pains not to disseminate, not to spread this allegation,' 'The charge was a serious specter that had to be resolved;' (she) 'proceeded quietly, but uncovered nothing.' She believed that their two proposals were turned down for other reasons. . . . Gonzalez was 'relieved' that it was rejected on its merits and that the CIA issue did not have to be raised."[28]

In a letter to Dr. Richard Nicholson, Acting Deputy Director, dated October 3, 1983, Gonzalez later informed the NSF that the CIA matter was not discussed during the panel's "actual" evaluation of the proposal, but instead while they were having lunch, which had been brought into their meeting room. Her letter recalled that she would not "brook" any consideration of the CIA business during the actual evaluation. For this reason, wrote Gonzalez, the "notes" of what the panel discussed did not reflect the lunchtime conversation. Although she couldn't remember exactly who said what, she was certain that everyone was "quite shocked" about how such allegations were "so casually" put into consideration. Furthermore, everyone was "distressed" that the allegations couldn't be shown to be true or false. She wrote that both over lunch and in the course of the entire two-day panel discussion, some panelists talked about "the fact" that Kalb and a "prominent" U.S. researcher appeared to have "bad feelings" about each other. Some panelists speculated that the rumor may have originated as a consequence.[29]

The preceding documentation debunks an NSF claim that a primary issue discussed at the panel meeting was "whether one of the RVRME researchers, Fred Wendorf, would be able to devote sufficient time to the research described in the proposals."[30] Although Redman's report indicates that one of the prepanel (outside) reviewers gave his proposal a "poor" rating, this was, wrote Redman, "the only pre-panel rating below excellent." Most telling, the NSF's own internal confirmation of Kalb's 13-point reconstruction of the panel discussion shows that the scientific merit of the Wendorf proposal "wasn't discussed in any substance" and that Wendorf's work in Egypt "was discussed and was not considered a problem."

The third proposal, whose principal investigator, Professor Bryan Patterson, was from Harvard, was considered by the Division of Environmental Biology (after being transferred for unknown reasons from the Systematic Biology Program). It too was rejected by the Foundation. Did the CIA rumor play any role in the discussions? According to the Foundation's own *Memorandum of Points and Authorities* (its court brief filed in response to Kalb's lawsuit, "It [the CIA rumor] was discussed on the Environmental Biology Panel."

Berkeley archeologist Desmond Clark had been asked to review the Patterson proposal; Clark also emphasized the CIA allegations. In Kalb's court complaint against the NSF, Kalb's attorneys pointed out that Clark had a grant proposal pending with NSF for anthropological research in southern Ethiopia, well removed from the RVRME study area. Kalb alleged, "Clark had previously told Gonzalez and other NSF officials that he was interested in using NSF funds to conduct the same research for which [Kalb] and his RVRME associates has unsuccessfully sought funds from NSF. In his review of the RVRME proposal, Clark again raised the allegation that [Kalb] was associated with the CIA and indicated that no scientist should work with [Kalb] until the allegation was resolved."[31]

After the Patterson proposal was rejected by NSF, a colleague gave Kalb a copy of Clark's review of the Harvard proposal. Here is what Clark wrote in his review of that proposal: "The qualifications of the senior personnel are beyond question and this reviewer is fully in sympathy with Mr. Kalb's efforts and energy in getting the RVRME together. However, his possible involvement with the CIA must, until this has been cleared up, remain a matter of concern for any scientists."[32]

How is it possible to know the review was written by Professor Clark? Clark's identity is revealed in the review when he refers to Fred Wendorf, whose proposal the NSF had already rejected: "It is Dr. Wendorf's intention to work these sites himself and he has asked this reviewer, who will be working again in localities to the south of the Middle Awash in early 1978, to join him in the survey, excavation and study of these sites. I have in principle agreed to do this with him in February and March, 1978."[33]

Wendorf identified Clark by name in a May 3, 1977, letter to Kalb, in which he informed Kalb that the SMU proposal had been turned down. "I was with Desmond Clark last Friday and took it upon myself to ask him if he could help out. I took this initiative because I knew you respected him highly, he has a project going in Ethiopia, and he is a damned good man. He was definitely interested, and we have tentatively agreed to a 'co-principal investigatorship.' If this goes through . . . I will join his group about mid-February. . . . All of this would require some deviation from Desmond's current research plans and some more money, but it is feasible should it be acceptable to you."

Clark was offering to pick up the pieces from the Wendorf proposal, which had just been rejected, according to the evidence discussed above, as a result of the CIA allegations raised by Clark's colleague at Berkeley, Glynn Isaac. "Clark's eventual move into what had been Kalb's territory make it all seem a bit collusive, even if all is innocent," said the NSF report later issued by Charles Redman.[34]

Wendorf made reference to the same conversation with Clark in a May 4, 1977, letter to Nancie Gonzalez. "I am enclosing a copy of my letter to Kalb which will give you some idea of how things are going and what seems to be in the wind with Desmond and I perhaps doing the Afar project together (with my participation and responsibilities very much reduced)."

Kalb, however, turned down Clark's participation in any research in which Kalb and his RVRME associates held the concession. Kalb felt that the mutual concern of Clark and Isaac in RVRME discoveries "was just a tad bit too cozy," as Kalb later described it in his Texas accent. Kalb informed the NSF of his position on Clark in a 29-page letter dated May 28, 1977, and he sent a similar decision to Wendorf on June 10, 1977.

On June 16, 1977, the NSF sent Professor Clark the Harvard proposal for review. Clark's review, stressing the CIA allegations, was received on July 28, several weeks after Kalb had, in letters to Wendorf and the NSF, refused him access to RVRME research territory. In a January 6, 1992, telephone interview, Professor Clark stated that he did not know at the time he reviewed the Harvard proposal that Kalb had turned down his participation in research in areas where the RVRME held the permit.

On August 15, 1977, three weeks after the NSF received his review of the Harvard proposal, Clark submitted to the NSF Anthropology Program (now under the direction of John Yellen) a proposal for the second installment of money from a grant that he had previously been awarded, NSF BNS76-83187, for study of Clark's own sites in southern Ethiopia. The proposed starting date for the additional money would be December 1, 1977. In his proposal, Clark requested that a substantial amount of money from this grant be shifted to work on sites that Kalb and his colleagues had discovered and for which they had unsuccessfully sought funds from the NSF. Clark sought to divert $47,000 to the "new" project from his own project funds. As a "continuing" award, the proposal for the new project would not have to go through peer review.

Clark's request was granted. After sending in a review tying Kalb to the CIA, Clark had convinced the NSF to let him use funds to work the very sites Kalb and Wendorf had requested funding for. On December 7, 1977, John Yellen, who had succeeded Gonzalez as Program Director wrote Clark, "I am happy to en close a copy of the Foundation's letter informing the University of the continuation of support for your project."[35] According to Kalb, Yellen had over the previous six months ignored repeated appeals by Kalb and Wendorf that NSF investigate the circumstances of the CIA allegations.

The NIH long ago structured its peer review system to avoid the potential problem of the anonymous reviewer with a conflict of interest. Applicants *know* who will be reviewing their proposals because the names of persons in each review panel, called "study sections," are published in advance. Although the applicants do not know which panelists have been assigned to report on individual applications, they can specify the panelists that they do not want to consider a particular proposal. Had this system been in place at the National Science Foundation, the potential conflict of interest in the Kalb case might have been avoided.

THE SCIENTIFIC WITCH HUNT

Upon tabling consideration of the SMU proposal, Nancie Gonzalez launched her own investigation to establish whether Kalb was or was not a CIA agent. Gonzalez later said in a letter to Dr. Richard S. Nicholson, Acting Deputy Director of the NSF, that in the course of her "probe" of the CIA allegations concerning Kalb, she had talked the matter over with both her own immediate supervisor, Dr. Richard Louttit, and lawyers in the General Counsel's Office. Both her supervisor and the legal counsels had "advised" her that her investigation was "appropriate."

How does an investigator go about obtaining from an espionage outfit a list of its agents? If real evidence had existed and been found, Kalb could, presumably, have been proven to be a spy. There is *no* evidence, however, that can prove that someone is *not* a spy. The insidiousness of accepting such a rumor is that everyone *except* a genuine CIA agent could be victimized by this kind of accusation.

On October 3, 1983, Gonzalez wrote to Dr. Nicholson that her main objective in conducting her investigation into the CIA allegations was to "protect" Kalb. In her view, Kalb had been vilified in violation of acceptable standards. Despite this, she acknowledged that many in anthropology were worried that the CIA was using field researching as a cover. Even though she considered herself patriotic, she was concerned about the danger to other American anthropologists who could be hampered or even prevented from doing their work if foreign governments began to distrust them as possible CIA agents. She believed that she was not in a position to immediately and perhaps improperly refuse further consideration to the allegations against Kalb.

Kalb's testimony to a congressional task force in 1986 indicates the result of Gonzalez's inquiry: "As might be expected, the Director could come up with no concrete 'evidence' that I was not connected with the CIA. In the course of her inquiries . . . questions were raised about a possible covert source of funds that had financed my 'activities' in Ethiopia, my 'connections' in Addis Ababa, and whether I had 'infiltrated' the Ethiopian government."[36]

"I'll give you an example of how far the CIA rumors went," Kalb said later. "It was claimed that the private U.S. foundation from which I received maybe $20,000, the Fund for Overseas Research Grants and Education, in Stamford Connecticut, was a CIA front. But the president of that organization was a former president of the American Association for the Advancement of Science, and one of the Board members was a Nobel prize winner on the faculty of Columbia University."

Kalb never received an opportunity to refute the accusations, a point the NSF conceded in its settlement with him: "At no time before or during the panel reviews was Mr. Kalb informed by NSF about the CIA rumor or given an opportunity to rebut it."[37]

In its apology to Kalb, the NSF stated: "NSF did not intend to give the rumor credence or to harm Mr. Kalb's reputation, or that of his American or Ethiopian colleagues by conveying the CIA rumor to the panels *and others contacted during and following the process of decisionmaking on the proposals.* NSF apologizes to Mr. Kalb if panel members *and others misinterpreted its actions* as an endorsement of the accuracy of the CIA rumor [emphasis added]. . . ."[38]

IMPACT ON SCIENCE

In the NSF's legal filing in response to Kalb's lawsuit, the NSF claimed "the proposals were declined for valid *scientific* reasons [emphasis added]."[39]

The Foundation's internal investigation into Kalb's 13-point letter, along with Redman's inquiries to former panel members, makes clear that the panels did not make any recommendations on the SMU and NYU proposals. They were withdrawn from the evaluation process while the Program Director conducted her off-the-record investigation. So the recommendation to decline funding seems to have been made by Program Director Gonzalez. (There are no documents regarding the NSF's decision on the Harvard proposal; the NSF claims to have lost the records.)

According to a March 12, 1990, letter from Charles H. Herz, the NSF's General Counsel to Eric R. Glitzenstein, of Public Citizen, who was Kalb's lawyer, the SMU proposal "was given the highest rating and was not funded only because NSF was later advised not to fund work in Ethiopia in light of the political situation there at the time." Yet, in 1987, the Foundation, in its legal response to Kalb's lawsuit stated: "In Anthropology, 'senior' proposals (as opposed to dissertation proposals) were reviewed and rated by both ad hoc reviewers and panels. The panels used an A to D rating system, with proposals rated A, B, or C being 'fundable." As a practical matter, however, low Bs and Cs were usually declined.

"The SMU proposal . . . was rated as a 'C.'"[40]

We know from the Redman report that there was very little substantive discussion of the Wendorf/SMU proposal. In fact, one panelist remembered its receiving an A rating. It seems as if the proposal must have been given a "C" rating by Program Director Gonzalez, *not the panel.* A "C," in essence, is a negative review. The outside peer review gave the proposal a much higher rating. "NSF does refer to five reviews for the SMU proposal," according to Kalb. "The principal investigator [Dr. Wendorf], however, was only given three reviews, which gave the proposal unanimously 'excellent' ratings."

The NYU proposal, according to the NSF in its legal filing, "was rated 'A contingent D' indicating a split on the panel on the fundability of the project."[41] However, the Redman Report, prepared for Vice President Bush's office, stated that the NYU proposal received "outstanding ratings." Furthermore, as noted earlier, the panel did not come to a conclusion since the proposal was tabled. The report of a "split" in the panel would seem to have come from the Program Director.

Charles Redman, in an October 29, 1982, memo, discussed the issue of the political situation in Ethiopia, vis-à-vis NSF funding, in this way. "I have been told that there was State Department (or at least NSF INT) [INT is the International Office] pressure not to send American researchers to Ethiopia at that time. I have seen no written record of those pressures in the jackets or files. Yet, this could have come via the telephone or personal conversations. To counteract this influence, Kalb sent Gonzalez a letter on June 30, 1977, with a statement from Aklilu Habte, the Ethiopian Minister of Culture and Sports."[42] Habte's June 28, 1977, letter to Kalb, which Kalb passed on to the NSF stated: "This is to acknowledge that the research of the Rift Valley Research Mission in Ethiopia is authorized by the Ethiopian Ministry of Culture and Sports Affairs through your contractual authorization and study area permit. This authorization applies to all those participants of your Mission.

"This Ministry endorses and supports the research of your Mission in the Afar Depression. We also strongly support your funding efforts on behalf of improving development of the National Museum, laboratory facilities, and the training of Ethiopians which we consider as a *sine quanon* for any research group that wants authorization and support from the Ministry of Culture to operate in the Country."

Charles Redman noted, "This letter is one of the strongest endorsements of a foreign mission I have ever seen from a third-world country. With much less than that in hand we send missions off to troubled areas."[43] Redman does state, "F. Clark Howell was also turned down that go around, which lends credence to the case that it was not the CIA allegations." However, Kalb claims the NSF funded one of Desmond Clark's graduate students to work in Ethiopia at this very same time, which the student did.

What was the impact of the NSF's decision not to fund Kalb's projects? "NSF awarded these researchers [Johanson, Clark, and Tim White, another Berkeley anthropologist] over $600,000 to perform the same research which plaintiff and his RVRME associates had intended to pursue with NSF funds," states Kalb's lawsuit against the NSF.[44] Kalb added in an interview, "Actually, NSF approved funds for this amount, but the researchers were actually awarded less since the Ethiopian government later put a stop to their fieldwork." *Science* summarized the results this way: "Berkeley scientists then took over [Kalb's] research grounds, with NSF support."[45]

Kalb stated before a congressional task force: "From late 1977 to late 1981, NSF's Anthropology Program embarked on a diplomatic and 'fund-raising' campaign—approved at the highest levels of NSF—to help secure a 'concession' from the Ethiopian government for the prehistoric sites previously studied by my organization, on behalf of 'American interests,' led by the same interests that had raised or promoted the CIA allegations with NSF. NSF's investment in this matter went so far as to give four consecutive *non*-peer review grant 'awards' to a senior scientist for this purpose."[46]

The senior scientist who received the nonpeer review grants was J. Desmond Clark, the Berkeley anthropologist who reviewed the Harvard proposal.

The Ethiopian government expelled Kalb and his family in August, 1978, the year after the Foundation turned down the grant proposals, while he and his colleagues were seeking money from other sources. "The CIA accusations were devastating, not only to me and my immediate colleagues, but to all those Ethiopians with whom we worked very closely." Kalb added in a letter to the author, "Many of my Ethiopian friends were harmed in various ways by the CIA allegations—those were dangerous days in Ethiopia; it was not child's play."

On November 6, 1978, less than four months after Kalb was expelled, Clark applied to John Yellen, the NSF's new Program Director in Anthropology, for a supplement to his original grant that had been funded for research in southern Ethiopia. Clark asked for an additional $11,611, to "assess the sites" left behind after Kalb's expulsion from Ethiopia. "Mr. Kalb's permit has now expired because, on August 1978, he was ordered by the Ethiopian authorities to leave the country. . . . it is especially expedient, for political reasons, to carry out this visit. At the moment, there is no other United States-based expedition working in Ethiopia and it is very desirable that American interests in paleo-anthropological research there should not be jeopardized as they would were such potentially crucial study areas to be given to other foreign-based scientists and/or expeditions."

Yellen agreed. On November 28, 1978, he wrote his recommendation for the nonpeer review supplement Clark requested: "Jon Kalb has been required to leave Ethiopia, and his research area is, in effect, up for grabs. This lends increased urgency to the supplement request, because it is important to the interests of U.S. science that American researchers maintain access to the extremely important Middle Pleistocene sites in the region. This supplement will meet that need. Given the remarkable success rate which Clark's proposals have had, the importance of the area within which Clark is working and the fact that Clark plans to resubmit a revised proposal this winter, this supplement will, in effect, also keep Clark's fieldwork going, albeit on a very much reduced level."

On September 10, 1979, Clark requested another $15,222 from NSF for a "seminar" on the Afar section of the Ethiopian Rift Valley—Kalb's RVRME study area. The new grant would allow Clark and his colleagues to visit Ethiopia for the purpose of persuading government officials to give Clark's Berkeley-based team a concession to the sites that had been discovered by the RVRME. For this effort, Clark was joined by Johanson and Yellen. Since the funds that a Clark sought were for a "seminar" (albeit an informal seminar at which no research papers were presented), once again the proposal did not require peer review. Yellen approved the fund request. At the same time, Yellen received NSF money for his own expenses on the Ethiopian trip. In addition, Clark requested funds to conduct a survey trip to the Middle Awash sites following the seminar. The purpose of the field trip was to view for the first time the extensive archeological discoveries that Kalb and his Ethiopian and American colleagues had made over three and a half years. Clark's fund request for the field visit was to be a supplement to the seminar proposal, so it too would bypass the peer review process. Yellen approved the additional award, making the fourth time the NSF had approved a grant to Clark without requiring peer review.

Maurice Taieb explained why Clark needed the reconnaissance visit: "I know everyone who researched in my original permit area, and Clark never did any work in the Middle Awash. Also, Johanson had never worked in the areas in which Kalb had the concession."

"On January 3 [1980], [Donald] Johanson and I flew to Paris en route to Addis Ababa," wrote Yellen in his internal NSF report on the seminar. He added, "The 'official' American group consisted of Drs. Desmond Clark, Don Johanson and myself." After viewing the Middle Awash region, Yellen described it as having "enormous paleoanthropological potential."[47]

POSSIBLE SOURCES OF THE CIA RUMOR

Johanson's presence at the seminar and field trip brings up an interesting footnote. Johanson always vehemently denied that he had originated the Kalb CIA rumors. One such public denial was broadcast on National Public Radio in 1982, on the program "All Things Considered," when he said, "It wasn't until this last visit, this year, 1982, that a number of [Ethiopian] officials said that he [Kalb] was expelled with 48-hour notice because he was considered to be a security risk to the country. And some people went as far as to say that it was almost certain that he was a CIA or they thought he was a CIA agent. And that is the first that I had heard those CIA allegations."[48] Yet, there is considerable testimony that Johanson knew of the CIA allegations much earlier.

How did the rumors start? Kalb wrote in a report to the NSF that in September 1974, following his falling out with Johanson, "[T]he Director of the EAI [Ethiopian Archeological Institute, an Ethiopian government agency] told me that Johanson had just alleged that I was a CIA agent. The Director related that during a meeting with him at the EAA, Johanson charged that I received covert funds through a CIA front organization, FORGE; also that I had a 'connection' with the American Embassy through a former research assistant whose father had worked with the Embassy. Johanson's argument was that no "known" foundation would support me because I lacked proper academic credentials and I was not associated with a bonafide research institution. . . . Based on the events that led to Johanson's allegations, the Director of the EAI told me that he gave the charges no credence whatever. . . ."[49]

Kalb was not the only one to suspect Johanson. When Dr. Wendorf wrote Kalb on May 3, 1977, to inform him that their proposal had been rejected, he stated, "[T]here was some question concerning your status. This last directly refers to gossip about you which stems from Johanson." Wendorf amended his comment in another letter to Kalb, dated May 25, 1977, "I should emphasize that the gossip did not directly emanate from Johanson but from one of the reviewers and also from one of the members of the panel. . . . To Johanson's credit, he refused to respond to a direct inquiry from the National Science Foundation on this matter." Fred Wendorf, the principal investigator of the SMU proposal, whose own career had suffered from the implications of the rumor, wrote to the NSF's then Program Director for Anthropology, Nancie Gonzalez, on May 4, 1977, "[T]he more I think of it the more I am convinced that Kalb is *not* a CIA agent. The ones I have known have never had that kind of scientific motivation, and he does have adequate, private means of support."

Five years later, Wendorf told Charles Redman that he "had heard the rumor himself directly from Don Johanson and from Glynn Isaac." On September 29, Redman recorded another interview, with another Kalb associate, Clifford Jolly of NYU, whose proposal had been rejected along with the SMU proposal. Redman wrote that Jolly "said Desmond spread the rumor in Addis. . . ."

Redman interviewed Johanson himself on October 17, 1982. "We are distant friends from graduate school, so this is quite informal," wrote Redman in his report, adding "in the course of this conversation he brought up Kalb. Johanson says the Ethiopian ministers still ask about him. The Swedish ambassador said that Ethiopian security threw Kalb out of the country for a CIA connection in 1978."[50]

In a 1991 interview with the author, Maurice Taieb said, "It may be that the CIA rumor, which personally I don't believe, was only used for scientific competition." Certainly competition for Kalb's site was fierce. In a June 1, 1975, letter to Taieb, Johanson wrote that, based on information Taieb had supplied him, the area covered by Kalb's permit extended "up to" the Hadar region, which is where the Lucy discovery was made. Johanson argued that he and Taieb should have full access to this area because they had conducted "so much" previous research there. They would find themselves "in big trouble" if Kalb got control of who could conduct scientific investigations in the area. Johanson urged Taieb to give careful thought to the consequences, and if he thought it necessary to travel to Ethiopia to talk with government officials about it, money could be provided for the plane ticket by the Leakey Foundation. Johanson emphasized the importance of the matter and that this trip just might be required.

Another paragraph emphasizes concern that Kalb's permit area would significantly reduce Johanson's intended research. Johanson stated that an Ethiopian official was under the impression that Taieb only wanted to conduct research in the Hadar region. Taieb must explain to this official that he and Johanson also require permission for the area south of Hadar parallel to the Awash, or else their research area will be "very, very, very small." Kalb has noted that the greater Hadar area is five and a half times the size of Olduvai Gorge, the area made famous through many years of work by the Leakeys.

Redman's 1982 internal NSF report on the Kalb case discusses Clark and Johanson's work at the Middle Awash sites discovered by the RVRME: "In many ways we have made special efforts to promote research in the region and certainly Johanson has had an important string of discoveries [at Hadar] and Clark is one of the distinguished archaeologists in the field. Yet, from a perspective of not being involved in the situation at the time, it seems to me that they have been given freer reign than normal projects and negative comments have been explained away too readily. Balance this against the unquestioned fact that some of the most newsworthy discoveries of anthropology in recent years have come from these researchers and perhaps they truly do deserve this special consideration."[51]

Redman's report was the result of an inquiry from Vice President Bush's office. The Foundation replied to the Vice President not by giving him a copy of Redman's report, but by giving him a memo prepared by the NSF's Director of Audit and Oversight, Dr. Jerome H. Fregeau. The memo stated, "There is no evidence that Mr. Kalb was in any way connected with the CIA nor that a rumor to that effect had any real influence on the decision, although, clearly, the existence of such a rumor was known." This is not at all the gist of Redman's Report, which had reached quite a different conclusion: "I do not consider the discussion of CIA affiliation at the panel meeting a mistake in itself, for surely it is relevant to our decision-making process. The problem I see is that if the proposal is rejected or tabled on that basis, then the issue must be clarified and if so the proposal should be granted or re-reviewed."

Fregeau's memo to Bush concluded, "The exceedingly cutthroat level of competition in East African anthropology is a longstanding problem. NSF cannot be blamed for it, but it must be kept in mind when making decisions on proposals in this area."[52]

On December 29, 1981, after J. Desmond Clark's first field season in former RVRME territory, he wrote John Yellen that in roughly 40 years of conducting field research in Africa, this was beyond a doubt and to a considerable degree the "most successful" research season of all. He noted that the sedimentary record appeared to be virtually all-inclusive for certain periods, ranging from the start of the Pliocene period all the way to the close of the Middle Pleistocene, and even a portion of the later Pleistocene. Not only were the fossils plentiful but they were also "superbly preserved."

Maurice Taieb said, "Tim White and Desmond Clark collected Kalb's reports to the Ethiopian authorities, but never gave credit to Kalb. . . . Desmond Clark found the opportunity to work in these areas from the work of Kalb. When the land was free and Kalb expelled, Desmond Clark could work more easily there." In his October 29, 1982, report, Redman referred to one of Kalb's collaborators, Clifford Jolly as saying that "the last Clark/White proposal relied heavily on the results of the Kalb mission, but did not invite any of them to participate."

THE NSF's SECRET FILING SYSTEM

Foundation Director Bloch mentions openness as a key element of the Foundation's peer review system. How open were the procedures in the NSF case examined here? When Kalb first met with NSF officials on January 23, 1978, to appeal the rejection of the RVRME-related proposals, he was told there was "nothing in NSF's files to indicate that the CIA rumor had played any role in the denial of the proposals, nor was there any record that reflected that the CIA allegations had been raised at the advisory meeting." When Kalb filed a series of requests under the Freedom of Information Act (FOIA) for what he believed were relevant documents to the decision, the Foundation replied that it couldn't find the documents. Eventually, over a period of years, Kalb did obtain some documents that he had requested, but not all of them; he received flat denials on the existence of other documents. According to Kalb, the NSF "withheld certain documents on the grounds that they are not in a 'system of records' covered by the Privacy Act, and that they can be withheld under [various] exemptions . . . of the FOIA."[53]

Kalb's concerns were confirmed by the NSF in a legal brief: "[B]ecause [a certain] memorandum was not kept in a file retrievable by Kalb's name in a group with other files retrievable by name, the Privacy Act amendment provisions do not apply."[54] Does this mean that the NSF simply couldn't locate certain documents because of some antiquated and inefficient computerized filing system? No, according to Kalb and the NSF Inspector General. The Foundation could easily get any document it had and wanted to get. The main proposal file stored documents by the name of institutions. Because the documents Kalb sought were ostensibly filed only under the name of the universities through which the proposals were submitted, the NSF denied his or the principal investigators' right to the records.

The peculiar arrangement was detailed in an April 1987 internal Foundation report that Kalb obtained through relentless FOIA requests. The report states: "The Foundation currently takes the position that portions of the grant proposal jackets [files], which often include names and comments of the peer reviewers, are not covered by the Privacy Act. . . . To support its position, NSF uses a hard copy which prevents direct retrieval of a proposal title by the name of the submitter of the proposal.

"In converting portions of proposals and grant jackets to electronic media, NSF must carefully design the new system to prevent retrieval by the name or personal identifier of the PI [Principal Investigator]. An adequately designed system will allow NSF to continue to withhold portions of the grant jacket."[55]

In other words, the NSF *deliberately* set up a filing system under institutional names to avoid supplying information legitimately requested by individuals under their own names, which the individuals specifically had a right to do under the Privacy Act. "A grant applicant may not know the full reason for the decline of a proposal if the agency is not complying fully with the access provisions of the Privacy Act," Kalb said in an interview with the author.

What Kalb was able to demonstrate to the NSF Inspector General was that the NSF had established an illegal "dual" filing system. The NSF claimed that proposal files were only retrieved by the names of a grant applicant's institution; however, the NSF could instantly locate through a cross-indexed computer system the name of the applicant and his or her institution. Such a dual filing system set up intentionally to circumvent the Privacy Act was a violation of the law. But the NSF's system was so clever that it was not detected for more than 10 years after the implementation of the Privacy Act in 1975. The NSF could easily have set up a procedure to withhold the identities of grant reviewers from applicants and thus protect the anonymity of the peer review process. But instead, the NSF chose to bypass the Privacy Act altogether by using a system that allowed the Foundation to *selectively* or *individually* withhold any document related to NSF decision making. The result, as Kalb discovered, was that for more than a decade the NSF filing system gave the NSF the ability to deny tens of thousands of grant applicants their legal rights of access to and amendment of records as allowed under the Privacy Act.

The double filing system was so obviously in violation of the law that the Foundation finally agreed, in 1990, following the investigation of their filing system by the Inspector General, to discontinue using it to withhold information.

As a result of these similar discoveries, Kalb and his lawyers at Public Citizen filed a petition with the NSF in 1989, asking, as allowed under the First Amendment, that the NSF change a number of its rules under which the Foundation operates. Shortly after the NSF received Kalb's petition, the Foundation announced a broad range of changes, incorporating almost everything Kalb and his lawyers had asked for. "It was no coincidence that these points NSF decided to make coincided point by point with our petition," said Kalb. Among the changes the NSF instituted were the following:

1. It will notify applicants that "they have Privacy Act rights with respect to their peer review."
2. Applicants will have Privacy Act rights to any communications from a reviewer concerning a grant proposal.
3. Panels will now keep accurate summaries that show not only what they decide but why.
4. All applicants will automatically get the panel summaries for their proposals.
5. The NSF will deal with conflict of interest by allowing applicants to review rosters of reviewers and designate those they do not want used.
6. The appeal process will be improved to include "substantive error."[56]

Will the new NSF procedures guarantee to eliminate future cases such as Kalb's? Perhaps. Still, the advance lists of reviewers that the Foundation will provide to applicants may contain thousands of names. And even if an applicant manages to spot an undesirable reviewer, the Foundation does not promise to avoid using that person. The conflict of interest issue could still arise, although its likelihood is, perhaps, lessened.

HAS THE FOUNDATION BULLIED ITS CONSTITUENCY?

"The word of mouth eventually spreads through the scientific community. People know when a proposal has been reviewed fairly," said Kalb in an interview with the author. How many others have felt that their proposals were reviewed fairly and yet have not gone to Kalb's lengths to try to get a fair hearing? Few have his persistence at getting at the facts. The NSF sent out a questionnaire to more than 14,000 campus-based individuals who were principal investigators of proposals sent in to the Foundation in the 1985 fiscal year. The NSF asked in the questionnaire whether or not they felt their proposals had been reviewed fairly. More than 9,500 respondents filled them in and sent them back, a staggeringly high rate of return for a questionnaire of this sort. Of the two thirds whose grant applications had been turned down, 60 percent, that is, 3,800 respondents, thought the decision on their proposals had not been made fairly. But the actual number turned down in 1985 was much higher than the 9,500 who returned surveys: 15,484 out of 23,605 proposals reviewed were turned down.[57] It is a reasonable assumption that the actual number who thought they had received unfair treatment was much higher.

Of those turned down that year, how many actually did something about it? The Inspector General's report said, "From . . . 1985 through 1989, the average annual number of requests for reconsideration was 37.8."[58]

What is reconsideration? The Inspector General says, "Reconsideration is a review of how NSF applied policies and procedures to ensure that the proposal was processed fairly. It is not intended to provide a forum for rebutting the program officer's judgment."[59] Any review doing that would be called an "appeal" procedure. As recently as 1989, according to the Inspector General, the Foundation "had no formal process to appeal a decision."

Kalb himself had requested reconsideration. Despite the known conflicts of interest and the unfounded CIA rumor, however, Kalb did not get reconsideration. "[The NSF] in fact *ignored* four separate reconsideration requests by me," said Kalb. A January 1978 memo signed by John Yellen and titled "Jon Kalb Reconsideration," states, "The three proposals to which Kalb refers were given a fair evaluation. They received proper handling and treatment and it does not appear to me that they should be given reconsideration."

In 1989, the last year for which data is available, 38 individuals wrote in for reconsideration. Exactly one received action. The Foundation reversed its decision to send it back without reviewing it. "The 38 requests for reconsideration . . . represent 0.2 percent of the approximately 20,000 declined and returned proposals [in 1989] . . . and indicate very limited use of the reconsideration process," the Inspector General stated.

After the conclusion of the NSF case, Kalb's attorney, Eric Glitzenstein of Public Citizen heard from other scientists who felt their proposals had been reviewed unfairly. Based on his conversations with them, Glitzenstein attempted to explain why so few go to complain:

"They're afraid. They believe there will be retribution, and that a much safer and wiser course of action would simply be to come back and play the game in the future, rather than risking having all of their future grants denied because they have complained about the process."

The NSF claims that it has changed its procedures and will now give scientists access to their files, although the reviews and other matters will still be anonymous. The Foundation also claims authors of rejected proposals can attempt to rebut earlier criticisms and rejections. Whether such procedures safeguard the scientists from the innuendo and conflicts of interest experienced by Kalb and his colleagues remains to be seen. But one point is clear—scientists must speak up when they feel their proposals have been treated unfairly. "Ultimately," said Kalb, "every scientist that individually and collectively fails to confront abuses and wrongdoing in the system is contributing to corruption in the system."

CHAPTER 2

Handing out the Big Money: Neither Science nor Sense

The previous chapter documented how the National Science Foundation (NSF) has decided cases concerning the funding of individual principal investigators, whose projects involve traditional, relatively small-scale science. This chapter will focus on questionable practices in the awarding of grants for "big" science projects—those small enough to go through the NSF's peer review process, but in which funding equals or exceeds $25 million, the congressionally accepted threshold for big science. I have again used a specific project to show in detail how the peer review process can break down without proper safeguards.

In 1986, the NSF was authorized to award $25 million to create an Earthquake Engineering Research Center. The question the NSF had to resolve was where to locate the facility.

According to the Congressional Research Service, California's seismic risk far exceeds that of any other state. The most populous state in the union, with more than 29 million people, California experienced 4,421 earthquakes registering 4 or more on the Richter scale between 1900 and 1986. Another 13,974 such quakes hit the neighboring states of Nevada, Oregon, Washington,

Hawaii, Idaho, Utah, Colorado, and Alaska. New York State, on the other hand, has experienced earthquakes of this size only 15 times during this century, and in the same period tremors of this magnitude occurred only 22 times in the combined states of New York, New Jersey, Pennsylvania, Massachusetts, Virginia, Connecticut, Maryland, Delaware, and Washington, D.C. Nonetheless, the National Science Foundation chose the State University of New York (SUNY)–Buffalo as the site for the Earthquake Engineering Research Center.

This chapter will examine the 1986 Earthquake Engineering Research Center decision in close detail because the documentation is more extensive than for other major decisions and provides an in-depth look at the actual behavior of the NSF's top officials and its top policy-making body, the National Science Board (NSB). My investigation is based in part on the NSF's written records.

Over the years there has been a gradual shift in funding of scientific research from smaller, more individual science projects to big science research. Rather than giving relatively small amounts of money directly to individual investigators, the NSF, with its merit review system, has increasingly given money in large chunks to a scientific center, which then dispenses it as it sees fit. The rationale behind this approach is that the recipient organization will then coordinate interdisciplinary research, focusing it on more complex problems, which the uncoordinated individual grants presumably are less likely to do. In addition, the new organizations supposedly encourage the business and government sectors to use the results of the coordinated research, that is, to "transfer" the new knowledge.

The award of the Earthquake Engineering Research Center to SUNY–Buffalo is one of a string of surprising big money decisions made by the NSF in recent years. In some cases, the NSF has canceled major grants even though its own peer review panels have recommended continued funding. In 1988, for example, the NSF shut down the Center for Robotic Systems and Microelectronics at the University of California–Santa Barbara, even though an NSF peer review panel had written, "It is the opinion of the panel that the successful implementation of the proposed center program would definitely be in the national interest. . . . The center funding should be renewed."[1] Another case, strikingly similar to that of the Earthquake Engineering Research Center, exploded onto the pages of the science press in 1990, when the

NSF awarded $120 million to Florida State University for a magnet laboratory. As the Florida State department is not a world leader in the field, it was a surprising choice. Two peer review panels had recommended the Massachusetts Institute of Technology (MIT), a recognized leader in the field, to house the magnet laboratory. The NSF's surprising decision led Massachusetts Senator John Kerry and Congressman Joe Kennedy to introduce language in the Foundation's funding authorization that requires the National Academy of Sciences to scrutinize the NSF's criteria in awarding grants and contracts and to investigate "the role played by outside scientists and executive agency staff."[2]

NSF decisions are not supposed to resemble the pork-barrel legislation sometimes enacted by Congress. In theory, they are based on "merit review," which the NSF claims is similar to peer review, with the addition of "other considerations."[3] As the awarding of the Earthquake Engineering Research Center illustrates, however, NSF "peer" review *can* be susceptible to conflicts of interest. In fact, the Earthquake Engineering Research Center decision raises questions about the very nature and practice of the NSF's peer review and merit review.

THE EARTHQUAKE ENGINEERING RESEARCH CENTER

According to the Committee on Earthquake Engineering of the National Research Council, the purpose of research in earthquake engineering is "to understand the nature of strong ground shaking and how to minimize its destructive effects."[4] The goal of the research is to enable improved construction and other earthquake safety measures.

Congress passed the Earthquake Hazards Reduction Act in 1977, earmarking money for earthquake research. Almost all the funds had initially gone to individual researchers as principal investigators in uncoordinated, individual grants. As part of a policy to set up research centers, top NSF officials believed an Earthquake Engineering Research Center would provide "a more systematic research approach dealing with . . . earthquake hazards." In November 1985, the NSF's engineering directorate invited proposals from universities for a special five-year grant of $25 million to set up and run an Earthquake Engineering

Research Center. Applicants would have to arrange for and show that they could obtain matching funds, dollar for dollar, from nonfederal sources—state, local, or private industry. The grant of $5 million per year would come from NSF money previously allocated for earthquake engineering research. The Research Center would use one third of the funds that previously had gone to individual earthquake researchers.[5]

The NSF's program announcement created an enormous controversy over one criterion used in judging the proposals—their national focus. The proposals were supposed to be judged according to the guidelines in the program announcement. According to a report prepared by the General Accounting Office (GAO), in 1987, the program announcement did not "explicitly state in either the criteria or introduction sections that the proposed center will be a national center and that it should have a national focus."[6]

The announcement does talk about "relevance to national technological problems," as well as "distribution of resources with respect to institutions and geographical areas."[7] But it is quite a stretch to make that a major criterion for deciding between proposals.

Nonetheless, the GAO reported, "[A]ll of the [peer review] panel members agreed that national focus was an important element in the proposed center." Indeed, the NSF had specifically told them to evaluate the proposals for this criterion. The NSF's Assistant Director for Engineering wrote, in a letter to Professor Joseph Penzien, Director for the Center proposed by the University of California (UC)–Berkeley, "The evaluation panel was instructed to evaluate each of the six proposals from a national point of view rather than as an eastern or western problem . . . the panel took into account the potential of national scope and impact. . . ."

The California team felt their proposal was evaluated on criteria that the NSF did not set forth in the original project description. As the GAO wrote, "NSF's instruction to the panel as they started their evaluation, we think, could suggest that national focus was a criterion added during the panel's deliberations."[8]

Penzien, on August 21, 1986, wrote to the NSF's Associate Director for Engineering: "[T]he Panel Chairman asked . . . how the California group would respond to proposals from other states. The California proposal did not include consideration of proposals for support from institutions in other states because this was not part of the original requirement. Upon completion of this discussion, the Chairman said something to the effect that that was what he needed to know. . . ."

In the same letter, the California group claimed their research would have national applicability: "California-based research has had major, if not principal, impacts on Eastern seismic codes, standards, and professional practices for decades. . . . The panel, not having such knowledge and experience, appears to have acted under the misconception that Eastern earthquake engineering problems are unique, and they obviously felt that these concerns would be better served by the New York proposal than by others."

Six universities had submitted proposals to the NSF for the proposed Center. Four of the universities had been eliminated on the basis of reviews of their proposals: the University of Missouri–Rolla, The Citadel, the University of Michigan, and the University of Illinois. The two finalists in contention for the proposed Center were The University of California (UC)–Berkeley and SUNY–Buffalo.

The consortium of California universities, led by UC–Berkeley, advocated a narrow focus for the research center, but one with crucial, practical applications—how to make existing structures safer during an earthquake. The consortium of universities headed by SUNY–Buffalo proposed something much broader. The GAO described it as "encompassing the entire spectrum of earthquake engineering."

In essence, Buffalo seemed to be saying, why limit earthquake engineering research to existing structures? Yet, as the Committee on Earthquake Engineering of the National Research Council has pointed out: "The estimated $8 billion of damage caused by the [October 17, 1989, California] earthquake was almost entirely to older buildings, bridges and homes. Such older structures and facilities have often been identified as providing the greatest earthquake hazard to U.S. cities, in the eastern . . . as well as the western United States. . . ." According to the Council, although construction in the United States costs over $400 billion per year, *existing buildings and facilities* are valued at over $10,000 billion, and most of them are not adequately resistant to earthquakes.

In defense of the NSF decision to award the Center to SUNY–Buffalo, Erich Bloch, then Director of the National Science Foundation in Washington, stated, "I do not think there is anything wrong with broadening the community researchers. I think the worst thing we can do is depend on one community only, isolated, by itself."[9] Bloch's objectivity is questionable, however, given that he is a Buffalo alumnus, although he disqualified himself from the selection process because of his Buffalo connection.

The "one community" the Director referred to, of course, was the California earthquake engineering researchers. "The Buffalo Center . . . has as one of its jobs to make sure that the transfer of information takes place between the research and the user end of it," testified Bloch. He also pointed out, "We did not put the center where the earthquakes are, but where the knowledge is, the interest. . . ."[10]

Many experts do not agree with Bloch's claim that the NSF put the Center where the knowledge is. Among these is a man who knows a great deal more than Bloch about earthquake engineering research and practice, Dr. Emilio Rosenblueth, Professor of Engineering, National University of Mexico. A world authority, Rosenblueth is a Foreign Associate of the U.S. National Academy of Sciences and the National Academy of Engineering and has a number of honorary degrees and distinguished awards. The Executive Committee of the Buffalo Center asked him to serve on the Center's Scientific Advisory Committee, after the NSF's award had been made. Here is what Rosenblueth wrote about U.S. earthquake engineering research in an October 17, 1986, letter to Thomas Tobin, Executive Director of the California Seismic Commission: "There is no place in the Western Hemisphere where better or more extensive research has been done and is being done than in the State of California. California has the most competent and committed people. This applies to the University of California, Berkeley, to the California Institute of Technology, and to Stanford University. . . ."

Rosenblueth's comments were echoed by the Federal Emergency Management Agency (FEMA). "Without question," said a FEMA Report, "California had the greatest concentration of eminent earthquake engineers at its major universities."[11] The state's then junior senator, Pete Wilson, elaborated on this when he compared California's research capabilities with those of New York: "The California universities involved in the proposed center are rated by the American Council on Education as among the very best of all civil engineering schools in the nation. In the same rating, SUNY–Buffalo is rated well below the California schools. UC–Berkeley is rated number one in the nation. . . . Even with an increase of 12 faculty members, the New York center will still have fewer faculty active in earthquake engineering than the more than 50 faculty members presently involved in earthquake engineering, education and research at the California universities participating in the California proposal."[12] Nor is the expertise limited to the universities. "Not surprisingly," reported the Congressional Research Service (CRS), "California is the State that is the most active on the state and local level in establishing programs to reduce earthquake hazards."[13]

During an interview on November 16, 1990, Bill Iwan stated, "California is the only state in the country with a State Seismic Commission." Iwan, who participated in the proposal, was head of the California Seismic Commission at the time of these events, and a professor of earthquake engineering at the California Institute of Technology (Cal Tech). He added, "The State has even developed a five year plan, 'California at Risk,' which is continuously updated, and is precisely for putting research into practice. No other state has anything like it."

Yet Dr. Norman C. Rasmussen, speaking before the National Science Board, which finally approved the SUNY–Buffalo award, said, "In my judgment, the others can say what they like, the technology transfer aspect of the proposal was very strong in Buffalo. That is, they thought about this and what they would have to do, and had some good ideas as to how to approach that problem and get the technology used."[14]

The verbatim transcript of the National Science Board meeting reveals no one informed the Board that California already had *in place* the nation's only existing governmental system for immediately transferring the knowledge. When Rasmussen asked NSF staffer Nam Suh, the assistant director for engineering, if he could supply any information along those lines, Suh said: "[I]n the case of Buffalo, one of the things they proposed to do was to establish information dissemination groups within the Center. So the research results they generated as well as the research results generated elsewhere can be supplied to the practicing engineers, as well as people who write codes, and so on and so forth.

"Contrary to that situation is the case in California, at the University of California. . . . The California people decided not to include . . . [dissemination groups] as part of the Center."

In fact, California already had an existing governmental mechanism for transferring new knowledge; they also had key personnel serving on both their Seismic Safety Commission and in their Center proposal. In addition, said a principal in the California proposal, "We emphasized bringing professionals into the Center to speed up the transfer. The peer review comments criticized our plan to transfer knowledge. I don't know how they could have done that."[15]

WAS THE DECISION FAIR?

Was the peer review that awarded the $25 million funding to SUNY–Buffalo a fair, unbiased decision? To answer that question, it is important to examine who was on the review panel. According to the GAO, the "NSF didn't want panelists from the states or schools submitting proposals." Although the intention may have been high minded, the effect was the exclusion from the panel of all earthquake engineers from California—in other words, an overwhelming number of top-flight earthquake engineers in the country. Curiously enough, no earthquake engineering experts at all were selected from any of the states west of the Rockies—a region that experienced the huge majority (18,681) of major earthquakes this century. Of the panelists selected, one was from Georgia (which has experienced 4 earthquakes of four or more on the Richter Scale this century), a second from Ohio (14 earthquakes this century), a third from Texas (14 earthquakes this century), a fourth from Massachusetts (1 earthquake this century), a fifth from Michigan (3 earthquakes this century), a sixth from Tennessee (19 earthquakes this century), and a seventh from Connecticut (no earthquakes this century). Four of the seven panelists came from corporations located in the East: John Rydz of the Emhart Corp (and prior to that the Singer Company); Dr. Constantine Papadakis at the time from the University of Cincinnati, but who had previously been with Bechtel; Dr. Mounir M. Kamal of General Motors Research Laboratories; Dr. James Beavers of Martin Marietta Energy Systems.[16]

program announcement. . . ."[28] The documentation referred to consists of notes made by the panelists when they reviewed the proposals and made on-site visits at the universities. These notes were then turned into reports, but the panelists did not write them. "The NSF staff, who were also present at . . . [the proposal reviews and site visits] prepared these reports on the basis of the notes supplied to them by the panel," wrote the GAO.

In other words, the NSF staff picked the panel and later wrote up the reports from the panel's notes. And there seems to be evidence of bias in the reports. For example, the panel's site report for UC–Berkeley says, "The present staff of the California team is excellent technically and well recognized but is aging,"[29] which would seem to imply that the California experts were too old. Age discrimination, however, is a violation of public policy, and the GAO stated, "We believe that it is not appropriate to mention age, and that on the basis of the data provided to us by both schools, this statement is misleading."

The age criticism was repeated behind closed doors at the meeting of the National Science Board. NSF's assistant director for engineering, Dr. Nam Suh, said: "[O]ne has to remember the fact that in terms of who's going to carry out the research five years from now, it turns out that, in the case of California, we have a great deal of concern because a large number of key people in the group are over 60 years old.

"In fact, one of the key members of the group is 73 years old. So, five years from now who is going to carry on the research burden? That's a question that the California people had to answer, and they could not provide an answer."[30]

The Californians protested to Senator Wilson, "This is totally false and misleading as the age issue was never raised with the California researchers."[31]

Suh wasn't the only one making remarks about the age of the California researchers at the NSB meeting. Dr. Norm Rasmussen, a board member, did so, too, when recommending the SUNY–Buffalo decision at the meeting: "California had a long-established program. I'm not an earthquake gut [*sic,* nut?] but from what I know they have better names in the field. They're also advanced in years."[32]

What were the ages of the California experts? George Housner, Professor Emeritus at Cal Tech, who was 76 at the time of these events, was undoubtedly one of the researchers who Rasmussen

could have been thinking about. The NSB members seemed to view Housner's age as a salient point. However, both the governor of California and the National Research Council, judging by their actions, attached much more importance to Housner's expertise than to his age. Three years after the SUNY–Buffalo decision the governor appointed Housner chairman of the Governor's Board of Inquiry for the October 17, 1989, Loma Prieta Earthquake. During the same period the National Research Council made Housner the chairman of the Committee on Earthquake Engineering.

Another person who was a likely target of the age comments was the leader of the UC–Berkeley proposal, Professor Penzien, who was 62 at the time of these events—hardly an advanced age. As for the other members of the California team, their ages were well distributed, with one age group ready to step into the shoes of older participants. Fifteen members were between 60 and 76 years old, eighteen between 50 and 59, twelve between 40 and 49, seventeen between 30 and 39, and one member was under 30.[33]

A second example of bias, or perhaps simple error, in the panel's report on the UC–Berkeley proposal was the comment, "the leadership is not firm in this proposal: a director has not even been named by the Chancellor." Perhaps the panel did not read the California proposal with sufficient care, because it clearly stated, "The first Director is Professor Joseph Penzien of the Department of Civil Engineering, subject to confirmation by the Chancellor."[34] The chancellor, of course, could not officially name Penzien to the job until the job actually existed.

Another possible example of bias can be observed by comparing the comments in the panel's report on the SUNY–Buffalo proposal and site inspection with the comments made about the California proposal and site inspection. Regarding Buffalo, the panel optimistically lauded its plans:

"It is believed that . . . major breakthroughs will result."

"The Center investigators also believe that . . . [efforts] . . . can be integrated . . . to handle the infrastructure problem."

"They are hopeful that . . . cost-effective seismic design and rehabilitation techniques . . . can be coordinated with other programs . . . concerned with . . . the nation's infrastructure."

"They are also helpful that . . . programs that investigate other hazards . . . can benefit from the Center's research on earthquakes as well."[35]

In contrast, the California report lacks any attempts at such optimism. Nor does the report compare California's first-class earthquake engineering expertise with Buffalo's enthusiastic aspiration to one day be first class.

The GAO investigators acknowledged the evidence of bias, but did not pursue the cause. In addition to the areas already noted, the GAO pointed out the following additional slanted coverage in the reports:

Corporate Representation. The New York site visit report stated, "corporate interest is very high," pointing out that representatives from industry were there during the panel's visit to Buffalo—apparently leading the panelists to think corporate cash would pour in. Yet, according to California's Senator Wilson, "nationally prominent" corporate earthquake engineers from California's "much larger and more active private earthquake sector" were present at the California site visit. And they too indicated they would make contributions. Did the California site visit report mention this? The Senator said no. "We agree," said the GAO.[36]

Implications of the Research. The New York site visit report claimed, "Many of the principles uncovered in its research program will have implications for earthquake hazard reduction in California and other Western states." In contrast, the report on the California site visit stated, "Although the research plan was focused on a national need it was very narrow and was directed only to the State of California." The GAO, in evaluating the reports replied, "We have no information as to why the panel believed that principles uncovered in New York's initial efforts would have implications for hazard reduction in western states even though it perceived regional differences in earthquake problems."[37]

Library. The New York site visit report asserts, "A library collection of documents, graphic materials and computer programs related to earthquake hazard reduction will be developed." California universities, of course, already had accumulated these materials over the decades and made them available to researchers around the country. Did the California site visit report mention these extraordinary library resources? No. Did the GAO feel that a discrepancy of coverage existed in the two

reports? The GAO stated, "We agree but did not obtain any further information relevant to this concern."[38]

Research Laboratories. SUNY–Buffalo had built a seismic research laboratory, and the New York site visit report touted it quite a bit. But the report on the California site visit didn't talk about the much bigger laboratories in California, except for the begrudging comment that they "would support a national earthquake center." Again, the GAO said about the discrepancy between the two reports: "We agree . . . and . . . this serves as another example of imbalance of coverage in the reports."[39]

The GAO summarizes these site visit reports in this way: "Overall, the reports present no negative evaluation comments regarding the New York proposal but, conversely, present almost no positive ones for the California proposal. . . . The tone of the reports in several cases was strongly positive for the New York proposal and as strongly negative for the California one."[40]

How could scientists, who presumably believe in objective, balanced reporting, author such biased reports? The GAO asked the panelists essentially that question. "[T]wo members told us that the reports do not reflect the thoroughness of their discussions and that, in hindsight, their deliberations should have been better documented," said the GAO.[41] Of the other five panelists, the GAO report says, "[S]ome stated that the reports were not meant to be a comprehensive explanation of the strengths and weaknesses of the proposals but to reflect the impressions of the panel." In other words, the official site visit reports advocating the funding of a five-year, $5-million a year award were to be considered as mere "impressions."

The GAO's report then becomes even more alarming. "Other members also stated that specifically in regard to the California site visit report, the panel (after their site visit to California and after making their final recommendation to New York) was trying to justify what they knew would be a controversial decision and that this could account for the negative tone." For at least some members of the panel, in other words, the bias or imbalance in the reports was *deliberate.*

The NSF did its own in-house investigation into the biased reports. The Division of Audit and Oversight, in a November 10, 1986, memorandum, seemed to find them acceptable.

EXAMINING THE DECISION IN DETAIL

Management Issues

One of the things the peer review panel was very concerned with in evaluating the two proposals was management of the Center.

The GAO reported: "The panelists told us that they were particularly impressed with Dr. Ketter's management abilities. Dr. Ketter's experience as a former president of SUNY–Buffalo and the impression he gave of being firmly in control at the site visit were key factors leading to several panelists' perceptions of Ketter as a stronger manager. Several panelists had reservations about Dr. Penzien's management experience. . . ."[42]

Apparently the NSF's Assistant Director for Engineering had reservations as well. "Just to collect all the best quarterbacks and then put them together as one team does not make that team strong. You have to have a team effort, you have to have a strong coach."[43]

Yet Buffalo's Dr. Ketter had a mixed track record at Buffalo. In 1967, as a vice president for facilities planning, he had put together a national committee of high-powered civil engineers to advise him at SUNY–Buffalo. (One member of that committee was a world-renowned earthquake engineer, Joe Penzien, who later was designated the Director of the proposed UC–Berkeley Center.) The major recommendation was to set up a big seismic simulator—a shaking table and other testing apparatus. When Ketter became president, he rounded up the money for these projects, which ultimately came to $2.25 million, but he could not find the faculty to utilize them. Many at Buffalo thought this collection of earthquake paraphernalia was an "underutilized white elephant."[44] Ketter later resigned as president and went back to teaching, with orders from the Dean to find faculty capable of making full use of the machines.

Penzien had never been a university president. On the other hand, he didn't have any underutilized white elephants in his closet, either. However, he did have experience getting high-quality work from his fellow scientists and was the founding director of Berkeley's Earthquake Engineering Research Center. Nonetheless, according to the GAO Report: the review panelists "perceived the New York researchers as having stronger experience and competence in managing a large organization or in

working together effectively as a team. In contrast, individual panelists perceived the California researchers as being inexperienced in working as a team, unsure as to the specifics of their management plan, or generally unconvincing as to their management abilities."[45]

The discussion of Ketter's alleged managerial prowess even found its way to the NSB's closed door hearing. Dr. Rasmussen, who played a key role on a last minute ad hoc committee recommending Buffalo, said, "The Buffalo program had a couple of very strong points. One was the former president of the university is the head of this project. He retired as president and chose—he was a structural man—to run this, and he's a dynamic, vigorous guy. Obviously a good manager."[46] Rasmussen didn't disparage Penzien, he simply didn't mention him.

The peer review panelists, with one exception, were not experts on earthquake engineering, but they were not experts on management, either. Any standard management textbook would point out that there is no one best management style. Management consultants and researchers may disagree on exactly which style is best in which circumstance, but they do agree, as one management textbook has phrased it, that "no universally accepted theory of leadership has been developed." Management experts are nearly unanimous in saying that management is not something where one size fits all. "Whether a decision should be made autocratically or democratically *depends on* characteristics of the leader, the subordinates, and the situation," says one management textbook.[47] Another authority states, "It is generally inaccurate to speak of effective and ineffective leaders since there is no such thing as a perfect leader for every situation."[48] Ketter may well have been perfect for the start-up phase of a new program at SUNY–Buffalo, which did not have the luminaries that the California universities had and still have. But a strong manager might have been a catastrophe with the UC–Berkeley group, whose individuals had been working together with worldwide acclaim for years and who seemed universally to agree that Penzien's style was the best for leading them. He had produced proven results as a manager of exactly the sort of people who would work in the proposed California Center in Berkeley. The FEMA Report explained: "The Californians had grown up professionally and had been successful under the NSF tradition that favored individual, competitive awards. . . . Although Penzien might be permitted to take the

lead on a given project, his colleagues were by no means 'followers.' They had status in their own right."[49]

As the GAO noted, "[T]hese distinctions [in management] were based more on the site visits than on the proposals themselves."[50] The panelists, that is, based their management judgments not so much on what the two teams had put down on paper, but on sizing them up in person as potential managers. The panelists aren't alone in putting so much emphasis on taking the personal measure of someone. Research on procedures for job selection routinely shows that most people rate the personal interview as the most important criterion by far. "[I]t is *the* basic selection tool," says a standard textbook on personnel.[51] Based on this universally applied measure, the California team came off, as managers, as the lesser of the two groups. The California group, however, received a good deal less of an "interview."

The dates of the Buffalo site visit, listed on the report's title page, are "July 8–9, 1986." The California site visit report lists only the date "August 9, 1986." Buffalo thus received a two-day visit, whereas California was limited to one day. The difference, according to the GAO, was not that the panelists spent an extra day rummaging through earthquake engineering equipment or attending slide shows. The panelists spent an extra day because they attended a dinner "hosted by the President of SUNY–Buffalo on the evening preceding the actual site visit."[52] In other words, the panelists and the authors of the Buffalo proposal had a few drinks together, told jokes, and did the sort of informal sizing up that is common in most high- and medium-level job evaluations. Few companies would hire a top-level manager or division head (let alone a chief executive officer), without spending some informal time over lunch or dinner; that is often where people take real measure of each other.

Why didn't the California group host a dinner as well? From the records, it is clear they tried. According to one principal in the California proposal, a dinner was scheduled to take place at the U.C.–Berkeley Faculty Club. The GAO report states: "According to the NSF official in charge of the award, a similar invitation for dinner on the evening before the California site visit was *declined* [emphasis added] because there was little time to schedule events and because some panelists were arriving late and the panel felt it necessary to meet before the actual site visit began."[53] The Berkeley team didn't flunk the management evaluation process—rather the panelists failed to use the same selection test for Berkeley that they used for Buffalo.

Faculty Requirements

The California proposal was also criticized for not recommending the addition of new faculty. Here is part of the NSB verbatim transcript in which two members of a critically important ad hoc committee expressed their thoughts:

DR. JAMES DUDERSTADT: I think the absence of a commitment in the proposal on the part of the California institutions to add the new faculty suggests that they really are not anticipating it for that long a term.

DR. RITA COLWELL: I think it's important to keep in perspective the fellow who chaired the site team visit. It was conveyed to him that, in fact, they were going to cut back on this component, and this proposal was viewed as a means of maintaining their salaries and status quo in comparison to Buffalo, where you would have a new initiative, a new thrust, a new focus, a new activity, new people coming in and a commitment of 12 new positions; which meant that this was not just a passing fancy.[54]

When the authors of the UC–Berkeley proposal read the transcript of this closed door meeting, they were furious. In documents prepared for Senator Wilson they wrote: "The Chairman of the Review Panel, Dr. Thomas Stelson, misrepresented facts and acted improperly if, as repeated to the National Science Board, he told the ad hoc committee that it was conveyed to him that the California group was "going to cut back" on new faculty and that their proposal was viewed "as a means of maintaining their salaries and status quo." Such statements were never made by any member of the California team either in public or privately."[55]

The California researchers were particularly incensed by both the leadership and staffing issues. All of them said that the California universities didn't need to hire 12 new faculty because they already had world-renowned faculties. One prominent member of the team said recently: "Management was supposed to be what they [the Buffalo team] were good at, but New York never got organized for years. Ian Buckle just became their first technical director.

They needed a technical director because Ketter wasn't qualified." Berkeley hadn't needed a technical director; Penzien was one of the most technically qualified researchers in the country.

In 1989, the NSF evaluated the Buffalo Center's success: "[T]he establishment of NCEER [National Center for Earthquake Engineering Research] has not resulted in any appointments of well-known earthquake engineering experts to research or faculty positions at the core institutions during the first two years of operation. Only the recent appointment of a Deputy Director can be noted as a departure from the pattern. A 'National' center must attract outstanding individuals. . . ."[56]

This criticism is quite sharp considering that members of a National Science Board ad hoc committee recommended the SUNY–Buffalo proposal partially because of the commitment to hire top earthquake engineering researchers.

One California scientist said, "No major figures want to move out there." Another important California scientist reported that the technical directorship had been dangled in front of his nose, but he turned it down. The unwillingness to move to upstate New York is hardly surprising. Not only would the established researchers have to leave the temperate climate of California for the colder weather of Buffalo, they would also leave behind easy access to California's relatively huge commercial earthquake engineering industry, with its opportunities for consulting contracts.

Funding Issues

An additional concern of the peer review panel in making its decision on where to locate the proposed Center was whether California could obtain matching funds for the project. For a while there was a real question whether such funding had been secured. According to the Federal Emergency Management Agency, "The NSF schedule and the California budget cycle simply were not synchronized."[57] The NSF had made matching funds a condition of the grant (although some applicants felt the NSF would drop the matching funds requirement). SUNY–Buffalo easily obtained the first year's $5 million of matching funds from a state agency that had wisely been set up specifically to free large sums quickly for this sort of project.

The California group struggled for months to obtain funding, lobbying the state legislature and the governor. In the end, they came up with $4 million in cash and the rest in in-kind contributions from the various universities, which together totaled $5 million. Despite the difficulty, the California group did finally marshal the state's political process behind them; only one member of the California legislature voted against the bill providing them with the money. Penzien's letter to the NSF telling them the Berkeley group had the matching funds arrived on July 16, one day before the deadline.

According to the GAO, "NSF did allow in-kind funds as part of California's match." The California earthquake engineers, however, were questioned by the NSF about the in-kind match, which made them think that "their proposal was viewed negatively for having in-kind funds."[58] Bill Iwan of Cal Tech described some of this questioning in a July 15, 1986, memo, in which he recounted a phone conversation that day between himself and the NSF's Dr. M. P. Gaus: "Dr. Gaus expressed his opinion that the entire 5 million dollar matching funds be 'honest-to-goodness money.' When asked what constituted such money or who would be making that decision, Dr. Gaus indicated that this may be a question for the NSF legal office. However, he made it clear that if the State of California came up with only 4.5 million dollars in matching funds, NSF would 'not be interested.'" In material prepared for Senator Wilson, the California team wrote, "Dr. Michael Gaus initially refused to grant a California site visit on the grounds that the Review Panel had directed that only 'fully funded' proposals would be considered and it was judged by NSF that California had not come up with a full $5 million in cash. . . ."[59]

The impression that in-kind money was not acceptable is reinforced by the California site visit report: "In-kind contributions of $1,464,000 for the first year represent existing support of the four institutions and do not represent any new sources of support. Furthermore, there appears to be little discretion to the Center for any reprogramming of these costs. In addition, some portions of these costs do not appear to be allowable in accordance with normal federal practice."[60]

Joe Penzien recounted his experiences regarding matching funds in an August 21, 1986, letter to the NSF. According to the letter, after he informed the NSF that he had obtained the full amount of matching funds, the NSF program officer told him that only the cash portion, $4 million, and not the in-kind portion, would be counted. Since $5 million in cash was supposedly required, the NSF would not arrange a site visit. Penzien wrote, "No such constraint appears in the program announcement." His letter recounts how he pressed the matter and was told that the NSF could have awarded the Center to Buffalo on January 16 because New York had by then come up with the money. Penzien pressed further, and the NSF program officer agreed to take up the issue with the Panel chairman. But the program officer got back to Penzien to tell him that the Panel chairman would still not arrange a site visit to Berkeley. At this point, the California researchers went to the California congressional delegation for assistance in getting a site visit. Penzien wrote, "The NSF responded by reversing its earlier position and scheduled a site visit for August 9." Senator Wilson confirmed that he intervened to get the site visit.

Both New York and California only had one-year funding commitments. But the site visit report for Buffalo said the New York group "expects" the money to keep rolling in. The California site visit report stated, "future year funding is not assured." "The facts are the same but the interpretation clearly biased," said Senator Wilson.[61] The GAO, in reviewing the site reports, said of the funding issue, "We agree that this statement appears biased. Both site visit reports state that the respective groups had matching funds committed for the first year only. . . . It is unclear to us why one school's commitment seemed more assured to the panelists than the other's."[62]

Interestingly enough, less than two years after the award to SUNY–Buffalo, the Center had trouble obtaining continued funding from the New York legislature. In 1988, Buffalo obtained only $4 million from the state. "It will place the program in considerable jeopardy. We can do 'a' job. Can we do what we've promised to do? No, we can't," said SUNY–Buffalo President Steven B. Sample.[63] The next year, New York State only provided $2.5 million in matching funds. Clearly the review panel's assumption that Buffalo's funding was more assured in the long term than Berkeley's proved illusory.

Bias Revealed through Timing

Perhaps the most telling factor of all, however, regarding the decision to award the Earthquake Engineering Research Center to the SUNY–Buffalo team has to do with the panel's timing of the decision. What the UC–Berkeley team discovered was that the panel had already recommended Buffalo to the NSF *before* the California site visit.

"The review panel did, in fact, recommend funding Buffalo prior to making a site visit to California," the Assistant Director for Engineering wrote Joe Penzien on August 21, 1986. That recommendation was made on July 9th, "immediately following the New York site visit," according to the GAO. The California site visit took place one month later, on August 9, 1986. "This recommendation was contingent on California not submitting its matching funds commitment by July 17, 1986," wrote the GAO.[64] It was an interesting approach to a panel evaluation of a major scientific project.

Did the GAO think the panel was prejudiced when they went to Berkeley for the site visit? The GAO wrote: "We cannot conclude whether or not the panelists and the NSF staff approached the California site visit with an open mind. However, given that the site visit occurred after a conditional recommendation had already been made and that the visit concluded with an unbalanced report, it is understandable that there were doubts about the fairness of the decision."[65]

The recommendation of the review panel was accepted by the NSF two weeks before the California site visit. Panel recommendations go to the Director's Action Review Board (DARB), made up of top staff officials, which votes on the panel recommendation. If the DARB votes in favor of the recommendation, it is sent for final approval to the National Science Board. "The DARB reviewed and approved that conditional recommendation on July 24, 1986, even though California had obtained a matching fund commitment and a site visit had been decided on by that time," wrote the GAO.[66] According to the NSF, this was the first time the DARB had considered a conditional recommendation.

Why did first the review panel and then the DARB make up their minds before they had all the facts in hand? The review panel told the NSF they did not want to meet again, in case California was unable to come up with the matching funds. Dr. Charles C. Thiel Jr., who had been the director of the earthquake engineering function at the NSF from 1971 to 1979, felt the panel's explanation for prejudging the Buffalo proposal didn't wash. On July 10, 1987 he wrote to Gary Aldridge, a Senate aide, that the panel could easily have evaluated the technical merits of all six proposals, or even of only the two finalists. Then, on the basis of site visits, the panel could have ranked the proposals that were acceptable from a technical standpoint. The panel could then have recommended the award to "the highest ranking proposal that had the matching funds available."

Instead, the panel voted in favor of Buffalo, and the NSF staff wrote the glowing site visit report for SUNY–Buffalo described earlier.

The DARB, the next body in the NSF hierarchy to approve the recommendation, explained to the GAO they pushed the recommendation forward because they wanted to make sure that all the paperwork was complete in time for the NSB's last scheduled meeting for fiscal year 1986, which would be held August 14–15. Thiel, who had a great deal of experience within the NSF with such deadlines, thought this didn't ring true, either. In the same letter to Aldridge, he wrote that "every program officer" knew that the NSB meeting in August was the last one that could decide on an award for fiscal year 1987. Not only was this "widely known" but the August meeting as the last for the fiscal year was "a pattern of practice for many years." It is part of the published schedules, pointed out Thiel. He argued that since the dates were known, program officers and other management officials could have given the appropriate priority to the EERC proposals. With only six proposals, wrote Thiel, "not much time is required."

The NSF staff had blamed part of their time pressure on the September 1985 Mexico City earthquake. The GAO accepted their explanation that the quake resulted in "the receipt of a large number of proposals concerning . . . [it] . . . and the subsequent increase in program officer workload."[67] But Thiel didn't buy it. He wrote that the closing date for unsolicited proposals was in mid-March 1986. A panel would review the proposals, and the NSB did not have to review any of the awards, which would not be made until August in any case. So, concluded Thiel in his July 10, 1987, letter to Aldridge, "[T]he argument for priority need for processing these proposals is specious."

What if California had received a glowing site visit report? According to the NSF staff, there would have been plenty of time to redo the paperwork. Yet, as the GAO points out, "NSF's assumption that sending forward a conditional recommendation would save time must be, in turn, based on the assumption that the recommendation would not change." In a telephone interview on

December 3, 1990, Dr. Thiel said, "If I had brought forward a recommendation with a review that was incomplete, I would have expected to be fired, or at least suffer a substantial consequence."

Nonetheless, the NSF investigated itself on the tentative recommendation and the Director of the Division of Audit and Oversight, Jerome H. Fregeau, in an internal review dated November 10, 1986, found the recommendation method acceptable.

The National Science Board

At the time of the SUNY–Buffalo decision the 25-member National Science Board, NSF's top policy-making body, had to approve all annual grants greater than $500,000. The Board is composed of the director and 24 others from outside the Foundation. But before the full National Science Board votes on a proposal, a subcommittee of the Board, known as the Committee on Programs and Plans (CPP), first reviews it. Apparently the subcommittee felt it was being rushed on the EERC decision and that they didn't have enough time to read everything they were given. "Additionally," said the GAO, "CPP members raised questions regarding the California proposal as well." So the CPP proposed to the chairman of the National Science Board to table the matter until the following NSB meeting. This was, after all, a large sum of money, the decision was politically loaded, and there was some basis for controversy. Their request however, was not satisfactory to the NSB chairman. Although it was okay to limit the review panel to "experts" from the East, it apparently was not okay for the CPP committee composed largely of Californians to delay the panel's recommendation pending further data. As a result, the chairman bypassed the CPP and set up his own ad hoc committee. Not surprisingly, that committee concurred with the review panel's recommendation to award the Center to Buffalo.

How did the ad hoc committee arrive at its conclusion? Dr. Rasmussen explained to the National Science Board, as revealed in the verbatim transcript of the NSB meeting: "[W]e spent about an hour, an hour and a half, talking to, in private session, Thomas Stelson, who was Chairman of the Review Committee for the National Science Foundation who reviewed all these proposals . . . And then we read . . . the two proposals last night." In other words, the committee rushed through the same material the CPP had been given, with no reservations about the limited time available to consider the material.

At the August 15, 1986, meeting, only one NSB member abstained when all the other Board members voted to award the Center to SUNY–Buffalo.

ADDING IT ALL UP

As a result of the NSF and NSB's decision, the Earthquake Engineering Research Center was located in Buffalo. "Merit" review by the National Science Foundation seems to have brushed scientific merit aside. The GAO said, "California school officials questioned whether the assessment of their proposal had been a fair one based on the criteria. From only reading the reports, it did not appear so to California or to us."[68] The GAO reported that at least some of the panelists deliberately biased the reports, and panelists and NSF officials put in official statements that did not square with the facts.

"NSF's documentation does not support a conclusion that a fair and balanced evaluation of the proposals took place based on the criteria in the program announcement for the EERC," wrote the GAO.[69] The following discrepancies occurred in evaluating the proposals:

1. The unprecedented conditional recommendation for Buffalo, which the GAO said did not have to be made based on the time constraints.
2. The addition of national focus as a criterion during the evaluation process.
3. The clearly biased site visits, during which the panelists sat down to dinner with the Buffalo group, but wouldn't do so with the California group.
4. The necessity for congressional pressure to get a California site visit at all.
5. The derogatory treatment of California's matching funds.
6. NSF staff statements to the National Science Board that were later disputed by the California team as contrary to fact, as well as the NSF's omission of critically important facts.

7. The NSB chairman's decision to disregard the suggestion of the established subcommittee (the CPP) concerning the project and set up his own ad hoc committee.
8. The use of a "peer" review panel for the Research Center that, according to several experts, had no research peers on it, contained only one person who was a recognized practitioner, and was biased toward the eastern region of the country.

Obviously, the decision raises serious questions. Yet, the GAO, despite the evidence in its own report, concluded that it "found no evidence that showed that one proposal was intentionally favored over another for reasons other than it better met the stated criteria."[70]

How did they come to that conclusion? "Since no detailed account of the panel's meetings exists, we could only reconstruct the evaluation process through the testimony of the panel members themselves. . . ."[71]

In other words, after collecting all the evidence, the GAO interviewed the review panelists as to whether or not they had been fair. When they replied they had been fair, the GAO decided the decision-making process was fair.

Would Congress have openly voted to fund the Earthquake Engineering Research Center in Buffalo, ignoring the area of the country that is most at risk from earthquakes? Probably not—voting to locate the Center in Buffalo couldn't pass the politician's standard test, "How would it look as a headline in tomorrow morning's paper?" As Senator Wilson acidly notes, "In February [1988] the center held its first major conference to assess earthquake damage on the East Coast. This is not unlike calling a conference to study hurricane damage in Kansas or marine life in North Dakota."[72]

"I fear that this decision may, unfortunately, result in substantially higher losses of life and property in the next U.S. great earthquake than would have been the case," wrote the chairman of the California Seismic Safety Commission.[73] On October 17, 1989, three years after the 1986 NSF decision was announced, a major quake [magnitude 7.1 on the Richter scale] devastated parts of the Bay Area, killing 62 people, injuring 3,750 others, and causing $8 billion in property damage. A number of people were crushed to death in a collapsed Oakland double-decked freeway.

Oakland residents crawled through the concrete ruble to pull strangers to safety. In the aftermath, a small but noticeable portion of San Francisco had to be bulldozed due to structural damage. Of course, even if the Research Center had been awarded to California, the benefits of the research would not likely have been in time to reduce the losses from the Loma Prieta quake. However, the quake gave a dramatic warning of the clear and present danger from earthquakes California faces every day.

DIVIDING UP THE PIE

Senator Wilson wasn't happy about either the NSF's decision to put the Center in Buffalo or the conclusion in the GAO's report on how the decision came about. He demanded a report on what was going on in earthquake engineering in this country, and he inserted language in the October 1988 NSF authorization bill that required it. The report, prepared by the National Research Council of the National Academy of Sciences, revealed the exact effect of the NSF's SUNY–Buffalo decision on earthquake research funding.

In 1986, the National Science Foundation handed out $1.353 billion.[74] Of this total figure, about 75% went to various kinds of research projects at universities, including 10% toward setting up research centers. The idea behind funding such centers was to concentrate interdisciplinary research into particular areas.

Because the money for the Earthquake Engineering Research Center came from a fixed pie, earthquake engineering researchers in California were afraid their research would suffer if the Research Center were located in New York. In fact, Senator Hollings asked the question, "Can NSF ensure that individual grants will not be cut in order to fund centers?"

Bloch replied: "The [centers] will not be just another way to 'pass' funds to individual investigators, and, therefore, they will not simply be a layer intervening between NSF and individual scientists. . . ."[75]

At the closed door National Science Board meeting, Nam Suh said, "[I]t's not as if this establishment of a center at Buffalo would shut off the California people or any other group."[76]

Despite these claims, the evidence suggests that California researchers did suffer. The Committee on Earthquake Engineering of the National Research Council compared the average annual grants for earthquake engineering research that went to individuals at universities during the 1983–86 period, before the creation of the Center, with the average amount individuals at these same universities received from the NSF during 1987–89, after the Center was set up. (All the amounts are in unadjusted dollars, so the change, allowing for inflation, is even greater than it might seem.) UC–Berkeley lost $764,000, skidding from $1,967,000 to $1,204,000. Cal Tech's funding plummeted from $822,000 to $354,00, a loss of $468,000. The University of Southern California (USC) lost more than $100,000, falling from $641,000 to $539,000. Stanford suffered the least, with a deficit of only $20,000. The University of California–Los Angeles (UCLA) had its budget nearly wiped out; it lost $122,000, in a budget that slid from $210,000 to $88,000.

Senator Pete Wilson helped to summarize what happened to earthquake engineering researchers in his state:

"At UC–Irvine, a newly constructed structures laboratory is not being utilized due to lack of funding. At UCLA, seismic tests of bridge and bridge components have come to a halt. At UC–San Diego, an NSF-sponsored testing facility will operate only at forty percent capacity due to no real increases in federal funding. . . . At Cal Tech, all full-time professional staff have been eliminated at the earthquake engineering laboratory which is essential to the completion of structural testing and ground motion studies."[77]

The Committee on Earthquake Engineering of the National Research Council agreed with the thrust of Wilson's assessment. "The funding reductions which have occurred outside the Center program have significantly reduced support for graduate students and have curtailed research. The decrease in research funds since the 1983–86 period totals . . . $4.3 million [annually]."[78]

By contrast, the universities tied in with Buffalo realized a substantial increase in funding from the NSF. SUNY–Buffalo's budget went up from $166,000 to $1,139,000. Other places received funding for the first time. The City University of New York was given $39,000, and the Lamont–Doherty Geological Observatory pulled in $383,000. Rensselaer Polytechnique Institute's funding went from $93,000 to $222,000. Lehigh University's earthquake engineering research money rose from $163,000 to $325,000. The Ivy League also did well: Columbia's budget shot up from $79,000 to $219,000; Princeton's jumped from $236,000 to $734,000; and Cornell received an additional $26,000, for a total of $362,000, but was also the recipient of $918,000 from New York State. In this redistribution of funds, the states west of the Rockies, where more than 18,000 earthquakes had occurred in this century, lost a considerable amount of money. The National Research Council Committee compared the NSF money (given directly or filtered through the Research Center) that went to individuals at universities east of the Rockies and at universities west of the Rockies before and after the establishment of the Center. "Even in the pre-Center years, the larger share of the NSF university grants (53%) went to the east," writes the Committee. But the percentages became even more skewed after the creation of the Research Center. "Since the establishment of the Center," reports the Committee, "62% of the NSF university money is spent in the east, counting grants awarded through the Center."[79] That leaves those facing the greatest risk with only 38% of the National Science Foundation money.

The California earthquake engineering researchers never doubted what was at stake; it was one of the reasons they submitted a proposal in the first place. "We felt we had to go along or lose the funding," said Stanford earthquake engineer Haresh Shah.[80] The Report for the Federal Emergency Management Agency said the same thing: "[V]arious supporters . . . pointed out [to the Governor's staff and key legislators] that the ERC decision was a pivotal one, and California could lose a great deal of its then-current Federal research support in earthquake engineering if it did not win the award."[81] The Committee of Earthquake Engineering of the National Research Council wrote a telling punch line: "[T]he geographical distribution of the funding has been out of balance relative to the seismic threat, and this imbalance was accentuated by the reallocation of funds resulting from the establishment of the Center."[82]

Nearly all the earthquake research money received by California universities is awarded under the NSF's standard, national peer review system. In comparison, funding from the Buffalo Research Center is awarded "by an executive committee with input from NSF and a Scientific Advisory Committee. Unsolicited proposals are not accepted. After discussion with leading investigators from the Center's core institutions, specific assignments are made."[83]

Hence, a third of the money that the NSF had previously handed out by national peer review was transferred to a local authority that could disperse the money as scientific patronage. This caused the National Research Council Committee to ask, "[A]re the matching funds sufficient to offset disadvantages such as the lack of peer review and the decline in funding seen by researchers outside the Center's operation?"[84]

The Center was supposed to produce results immediately to help people counteract the destructive effects of earthquakes. How effective has the Center been? According to the most recent available information, the NSF's Third Year Review, the Center "appears to be making more progress than would have been possible with individually funded projects involving the same researchers and institutions.[85] In other words, if the same people, at the same institutions, had gotten the same money as traditional individual research grants, the results might not have been as good. The Committee on Earthquake Engineering of the National Research Council described the preceding sentence of the NSF's Third Year Review as "a rather weak endorsement of the Center."[86] Given the documented shift of NSF money from expert researchers to less qualified researchers, at least some of the researchers and institutions the Center funds undoubtedly would not have been funded under peer review.

At the time of the award to Buffalo, a Midwestern researcher was quoted in the *Los Angeles Times* as saying, "It is very likely that the development of new ideas will slow down for a while. There is no history of [earthquake] work at Buffalo; the government is making a leap of faith.[87] Nor has that leap of faith yet been justified, according to the NSF's own third year site evaluation of the Buffalo Center: "Technology transfer beyond that provided by information services, by distribution of ground motion data, or by development of computer software has yet to occur."[88]

AN APPRAISAL OF THE BUFFALO CENTER'S EFFECTIVENESS

Senator Wilson's rider to the 1988 NSF authorization bill required that the most prestigious group of scientists in the country, the National Academy of Sciences—whose principal operating agency is the National Research Council—conduct a study of earthquake engineering research efforts in the country. The chairman of the National Research Council's Committee on Earthquake Engineering was George Housner, a member of the National Academy of Sciences, who had been specifically alluded to during an NSB meeting as being too old.

Housner decided not to appoint a new committee to write the Report. "I thought the result would be better if we did it. You could hardly appoint a committee that wasn't involved one way or another with the Buffalo decision, and would appear biased. The existing committee had been appointed before the Buffalo decision."

Was the Committee biased against Buffalo? Apparently not. "Five of the eleven members of the Committee either were on the Buffalo Center's Scientific Advisory Committee, or the company at which they worked, or they themselves personally were getting grants from the Center," according to a California earthquake engineer knowledgeable on the matter, who asked not to be identified.

The report the committee approved contained criticism of the Buffalo Center. When the report was circulated to the various federal agencies concerned with reducing the damage from earthquakes, the NSF strongly objected, according to several persons familiar with the events (who asked not to be quoted). These persons stated that an NSF representative didn't want any negatives in the Report. The October 17, 1989, earthquake had just shaken the Bay Area, and the NSF staff person reportedly said that Congress might spring more money for earthquake engineering—provided the earthquake engineers weren't knocking each other in official reports.

There is independent corroboration for this in a June 22, 1990, memo by an official charged with analyzing all the reviews of the report from government officials prior to its official publication. The official wrote, "I have completed my review of the reviews and they have left me in a rather confused state. We have been asked to make a critical review . . . but nobody wants us to be critical. We are being told not to make waves because the golden goose might drown." He added that NSF officials "want us to remove the facts about funding patterns, especially the BC vs. AC [before Center versus after Center] comparisons. Are such facts really so boring to Senators and their staffs?"

For whatever reason, the Committee members deleted some of the findings and criticisms that I have quoted in this chapter. These extracts cannot all be found in the final report because they are from the 4/11/90 draft of the Report. After reviewing the Buffalo Center's published research papers, for example, the 4/11/90 draft of the report said "the technical quality is judged to be mixed," an assessment that seemed in line with the NSF's third-year site visit report, quoted earlier. This critical statement in the National Academy of Sciences Report, however, was dropped in later drafts. All the obvious judgmental statements were deleted. Even the statements quoted earlier concerning the redistribution of funding as a result of the center were cut, although the data on which they were based appeared in the tables, if anyone wanted to puzzle through them. Although Congress had specifically requested an "assessment of the adequacy of each agency's current Federal earthquake engineering efforts," the NSF and other government agencies had managed to kill every critical assessment in the report.

WHAT THE CASE MEANS TO SCIENCE FUNDING

The value of merit played a key role in this case. On the one hand, the best scientists and engineers, who had previously been rewarded for their excellence with faculty positions at the most distinguished universities, were not funded. On the other hand, people of lesser achievement were funded. Clearly merit, in this case, didn't pay.

Obviously, the meritorious, with short-circuited careers, were not the only ones watching. Word gets out on what pays and what doesn't. In this case, whom you know clearly was more important than what you know—precisely the reverse of what should prevail in science.

Should we have confidence in the judgment and procedures of NSF officials in more difficult cases after seeing how they operate in a relatively straightforward case?

In the Earthquake Engineering Research Center decision, many of the discrepancies were exposed by a General Accounting Office investigation ordered by Congress. But what has happened in cases where Congress has not mandated a full-scale investigation of the NSF? The case raises questions about the systems used by the Foundation to award science funding, especially their potential to be manipulated by political or other considerations.

Top NSF officials seem to have decided that they want to encourage big money awards. For example, in the two years preceding the Earthquake Engineering Research Center decision, the NSF set up 11 other engineering research centers. If large science projects are not free from political influence, how can we expect smaller, less visible projects to be decided fairly and objectively?

The NSF officials do not control all the science money in this country: The Department of Energy spends roughly three times as much; the National Institutes of Health passes out nearly four times as much; the Pentagon disposes of more than 20 times as much; and the private sector funds roughly as much as do all other sources combined. But outside of a few select fields, including certain areas of physics and nearly all of biomedical research, the NSF pays for the lion's share of basic science research done at U.S. universities. This chapter and the previous chapter raise disturbing concerns about the funding of such research—and the very health of the science system.

CHAPTER 3

SUPER SCIENCE AND POLITICS

Because congressmen and senators face reelection, they have to "deliver" for their constituencies. In addition to taking stands on ideological issues, such as abortion, legislators provide their constituencies with patronage. Unfortunately, congressionally funded "super science" projects often seem to become casualties of such patronage. In a telephone interview on July 20, 1990, Ernest Fitzgerald told me, "For the most part, the huge science programs are just big patronage programs; of course you have to have a scientific core of seeming necessity on which you can layer these huge blobs of fat in the form of excess costs," Fitzgerald, an Air Force cost cutter, did extensive research into the Space Shuttle for then Senator Proxmire. An investigator for the Congressional Research Service (CRS) put it more gently when he said, "[D]ecisions about the levels and purposes of specific public [super science] R&D expenditures will remain essentially incremental, judgmental, and political for the foreseeable future."[1]

An example is the "Superconducting Supercollider" (SSC), intended to be the world's largest high-energy particle accelerator. In a report to Congress, the CRS acknowledges that not everyone feels the Supercollider will be worth the billions it will cost. But the report also notes, "There would be economic benefits to the locality in which the SSC would be built."[2] Naturally, every congressman wanted the Supercollider in his district, and every senator in his state, a fact that Republican Congresswoman Claudine Schneider noted in a November 1987 article: "The chorus of support for the SSC is, indeed, a loud one. More than half the members of the House have signed on to the measure as cosponsors, an extraordinary showing of bipartisan unity. Still, their song has only begun. Once the Department of Energy chooses the site for the project, I believe that chorus will be reduced to a solo, or at best a duet.

"Twenty-five states are colliding in the selection process. A dozen others have endorsed proposals submitted by their neighbors. Not surprisingly, a whopping 94 percent of the cosponsors of the SSC bill represent one of these states. Such a lopsided margin draws into question the motives for supporting the supercollider. Is it, in fact, concern for America's long term future or, more likely, the short term prospect of thousands of new jobs and millions of dollars of added revenue for the winning congressional district?"[3]

When the site chosen for the Supercollider turned out to be Texas, the home of newly elected President Bush, in the Republican Sixth District (next to the district of then Democratic Speaker Jim Wright),[4] the senators from Mississippi, Arizona, Illinois, Michigan, and Colorado requested an investigation by the General Accounting Office (GAO) into the site selection process. In answer to the Mississippi senator's question whether the site selection process had been fair, the GAO said that it had been. But, added the GAO "for any future site selection process similar to the supercollider's"[5] the process could be improved. In other words, if the government ever got around to building another Superconducting Supercollider, perhaps things could be done differently.

The collider itself can only be put in one place, but the construction of its components can be spread around into many other congressional districts and states. The possibility of patronage, in the form of job-creating contracts, can help to turn opponents of super science projects into proponents. In April 1988 Senator J. Bennett Johnston of Louisiana, chairman of the Energy Committee, expressed skepticism of the then estimated $4.4 billion cost of the Superconducting Supercollider. "We now are in a position of wanting more than we can pay for," said the senator. The Appropriations Subcommittee on Energy and Water, which the senator also chairs, added in its fiscal 1989 report, "The committee simply doesn't know where the money is going to come from." Louisiana, however, is a poor state that needs jobs, and a good senator will do what he can to help out. General Dynamics informed Johnston that if the company got the contract for $1 billion or more to build the huge magnets for the project, the company would be happy to build them in a new plant in Louisiana. Johnston, in fact, made the announcement of General Dynamics' plan, just before his committee approved a $225-million funding bill for the SSC, which was $25 million above what the House was willing to budget for. Johnston said, "I've always been a supporter of high-energy physics." At that point Grumman jumped in and announced that it too would build a plant in Louisiana if it received the contract.[6]

THE PENTAGON CONNECTION

Both General Dynamics and Grumman, the main contractors bidding on the big magnet contracts for the Superconducting Supercollider are major Pentagon contractors. Many other Pentagon contractors are involved in building various pieces of the Collider, which has contracts in so many states it is referred to not as a pork barrel but as a "quark barrel." Those states include California, Oregon, Washington, Nevada, New Mexico, Texas, Missouri, Illinois, Wisconsin, Indiana, Ohio, Georgia, Florida, North Carolina, Virginia, Pennsylvania, New York, New Jersey, and Massachusetts. The Hubble Telescope and the Space Station, two other super science projects mandated by Congress in the past decade, show a similar list, approaching half the states in the country. Lucrative weapons contracts are parceled out in the same way.

Major Pentagon contractors often become involved in super science projects. The Hubble Telescope, costing $1.5 billion (plus another $200–$300 million operational costs) had only two main contractors: Lockheed, in Sunnyvale, California, and Hughes–Danbury, in Danbury, Connecticut; both are Pentagon contractors. (Hughes had bought the Danbury unit, then known as the Electro Optics Technology Division, from the Perkin–Elmer Corporation of Norwalk only about seven months before the Hubble

Telescope began orbiting the Earth. The unit also makes components for spy satellites for the Pentagon.) Pentagon contractors with Hubble subcontracts include Martin Marietta and Fairchild Industries.

Even the seemingly most benign super science projects, such as the Human Genome Project, have close ties to the military—in this case, the two major weapons laboratories run by the University of California, Lawrence Livermore and Los Alamos. Both are funded by the Department of Energy (DOE), and every scientist working in them has a top security Q-clearance. A CRS report explained how the connection between nuclear weapons labs and genetic mapping came about. "In the past, much of the genetic research supported by DOE was aimed at examining the mutagenic effect of environmental contaminants like radiation on human genetic material. As a result, DOE employs scientists who have expertise in disciplines like molecular biology, engineering, and mathematics."[7]

Civilian-oriented researchers at such labs can also end up collaborating with weapons scientists. Robert K. Moysis, the Director of the Los Alamos Center for Human Genome Studies told a senate committee how biologists end up working with scientists in fields far removed from the life sciences: "An amazing thing is happening because of the 'big science' Human Genome Project. Because of its complexity and scale, biolgoists at Los Alamos have found great advantage in working side-by-side with their colleagues in the physical, engineering, and computational sciences. This interaction has produced novel solutions to the problems of complexity and scale, including robotics, automation, and computational algorithms that have freed humans from the 'routine, boring' tasks the critics oppose."[8]

The Pentagon-linked participants on the project are spending Pentagon-sounding sums of money, which may ultimately run to billions of dollars. According to the CRS report, the DOE spent $10.7 million in fiscal year 1988 on human genome-related research, $18.5 million in 1989, $27.9 million in 1990, and an estimated $47.9 million in 1991. These figures actually understate the DOE outlays, because, according to the CRS report, they do "not include salaries and expenses of DOE employees devoted to this effort." Moreover, the DOE spends less than half of what is paid out by the National Institutes of Health on the Genome project.

Which companies receive super science money? The Star Wars (Strategic Defense Initiative; SDI) project shows a typical super science distribution. In fiscal year 1990 the SDI budget was about $3.6 billion, (half the projected figure of five years earlier, before the arms race with the Soviet Union ended). The money was spread among more than 1000 different contractors, including Lockheed, Boeing, TRW, Rockwell, General Motors (Hughes), McDonnell Douglas, Teledyne Brown, Martin Marietta, the Massachusetts Institute of Technology, and Raytheon. The shower of Star Wars gold fell on every state except Arkansas and North Dakota, ranging from California's $5.5 billion to Delaware's $101,000.[9]

There is one fundamental difference between smaller-scale Pentagon science projects and civilian super science projects. Although Pentagon projects may fail totally, they can still be put into mass production and the general public will never know of the projects' shortcomings. Those who do, and should care, such as military officials, usually keep quiet. But when a civilian super science project flops, it can't be hushed up. As we saw with the Space Shuttle Challenger catastrophe, too many people are watching. For example, after the National Aeronautics and Space Administration (NASA) launched the Hubble Telescope and its mirror was found to be defective, *Newsweek* splashed across its July 9, 1990 cover, "NASA's $1.5 billion blunder."

The mistake the government makes is in assuming that the Pentagon contractors' work will be better for the civilian super science projects than it has been for the Pentagon. "The major defense contractors . . . make money primarily by selling allowable costs to the government," said Ernest Fitzgerald in a memo to an Air Force general.[10] That is, contractors who work for the Pentagon get paid not for delivering workable goods, but for delivering costs that are reimbursed by the government, along with an ample prenegotiated profit. Fitzgerald added in an interview with me, "The contractors are always happy when the product turns out okay. If it doesn't, that's just one of the chances you take when you're working on the fringes of man's knowledge." The vice president of Perkin–Elmer, the prime contractor for the defective Hubble mirror, explained seven years before the telescope was launched why his company was so late in delivering and had spent so much over budget: "Our problems resulted from a combination of underestimating technical challenges and the cumulative effect of funding shortages. This funding shortfall in turn resulted in elimination of development testing, cutbacks on critical support hardware, interruption in certain development efforts and general operational inefficiencies."[11]

Fitzgerald did extensive research for Senator Proxmire on NASA pricing, particularly on the Space Shuttle. "Almost all government agencies use Pentagonal procurement practices. The procurement guide is government wide and the different agencies just adopt the armed service procurement manual and put a new cover on it," said Fitzgerald. He added, "That's the way the Pentagon has infected the whole government with these dreadful policies."

Some science officials have intimate knowledge of Pentagon procurement policies because they were Pentagon contractors themselves. In 1986, the nominal head of NASA, James M. Beggs, who had held the post since 1981, was on leave because he was under federal indictment for defrauding the Army on a prototype of an unworkable antiaircraft gun, the Sergeant York. Beggs and three fellow executives from his old company, General Dynamics, had been indicted in December 1985.[12] Charges were dropped against all four in June 1987 when Justice Department Assistant Attorney General William F. Weld said, "We were wrong." The Assistant Attorney General quickly added that it had been hard to believe that General Dynamics could collect $39 million and then "could deliver a bucket of bolts" but that is essentially the way the contract read provided the contractors put forth their "best efforts."[13] In 1989 NASA's Inspector General, after auditing NASA subcontracts, stated, "Subcontractors were paid an estimated $22 million in excessive profits. . . . About 81 percent of the subcontracts reviewed were awarded without competition. Approximately 33 percent of the subcontracts were overpriced and 12 percent had profits exceeding 100 percent, some reaching nearly 300 percent."[14]

As with Pentagon projects, government overseers on super science projects come and go. But prime contractors remain constant. For example, Bertram R. Bulkin, the Director of Scientific Space Programs for Lockheed Missiles and Space Company, Inc., testified, "I have been personally involved in the telescope program since 1973. I was the Hubble space telescope program manager through the year of 1985."[15]

By contrast, the government scientists working on these projects usually change several times over. Senator Gore noted, "NASA's problems in managing the Hubble become clear when one realizes that during its 12 years of development there were a total of 15 program managers between the Marshall and Goddard Space Flight Centers and NASA headquarters." The result is that NASA loses its institutional memory. The NASA personnel lack continuity and assume, with complete justification, no responsibility for the actions of predecessors. When the defective Hubble Space Telescope began orbiting the planet, Senator Gore asked then Associate Administrator of the Office of Space Science and Applications, Lennard A. Fisk, PhD, if he or any of his colleagues knew about a critical issue in the early 1980s—whether NASA could have used an existing Defense Department facility to test the big mirror. "I have no knowledge," replied the NASA official. The reason? "I think in all cases we are more recent additions to the program than the events that would have taken place in the 1980s," replied Dr. Fisk.

In addition, the NASA organization for the Hubble Space Telescope is fragmented between two major installations, Goddard and Marshall. "Could it be," asked Senator Gore, "that the loose and poorly coordinated management structure used by NASA could have resulted in more autonomy for Perkin–Elmer than was healthy?"[16]

Lack of managerial control is apparent with respect to the Hubble Telescope's big mirror. NASA asked the prime contractor for the mirror, Perkin–Elmer, to find a subcontractor to build a backup mirror, just in case Perkin–Elmer's was defective. The contractor hired Eastman Kodak, which had been the original competing bidder for the mirror contract, and whose mirror reportedly turned out to have some better properties than the one built by Perkin–Elmer. The head NASA scientist on the Hubble at the time, Dr. C. Robert O'Dell, now of Rice University, argued unsuccessfully with Perkin–Elmer to use the Kodak mirror. "They didn't want to use somebody else's mirror in their telescope," he explained. "Naturally, they found reason not to use the other guy's."[17] So contractor dominance guaranteed that the inferior mirror was used.

Current officials worry that contractor dominance may be getting even worse. James Thompson, Jr. NASA's Deputy Administrator, wrote the head of the Office of Management and the Budget, Richard Darman, that he was concerned that OMB was pushing to "convert more and more critical functions to contractors."[18] According to Senator Gore, "NASA's over-reliance on contractors is cutting back on NASA's inherent ability to manage things well itself."

LOW-BALLING CONGRESS—THE BUY IN

At their initiation, super science projects usually promise a great deal for relatively little money. The Hubble Telescope was originally pitched to Congress as the "large space telescope," with a sticker price of $435 million in 1978 and a range of possibilities that far exceeded its expected capabilities even before the discovery of the defective mirror. By then the project had been scaled down and was simply called the "space telescope." Among the things dropped from the project was end-to-end testing of the mirror arrangement."The end-to-end test was judged . . . to not have a good cost-to-benefit ratio" said Dr. Lennard Fisk, NASA's Associate Administrator for Space Science and Applications. He acknowledged that the project had "the largest cost overrun of any project we have ever had."[19]

The Superconducting Supercollider was originally pitched to Congress in 1986 for $4.4 billion, although the Central Design Group that made this estimate admitted even then that such extras as preoperating costs could run the bill to $5.9 billion. By mid-1990, the cost was estimated at $7.2 billion, and rising. By late 1990, the cost estimates ranged from $8.2 billion, (according to the Deputy Secretary of Energy), to $11.8 billion (according to a DOE Independent Cost Estimating panel).[20]

Precisely the same sort of escalation occurred with the Space Shuttle. In 1972 when the project was essentially finalized, it was supposed to have been able to put objects in space at a cost of $100 per pound. At these prices, economist Oskar Morgenstern, coauthor with John von Neumann of *The Theory of Games and Economic Behavior,* had told NASA that if the Shuttle flew at least 30 missions per year, it would be commercially viable, paying its own way. This number of flights was not expected to be a problem because NASA planned for the Shuttle to make 60 flights a year. In 1985, the Congressional Budget Office calculated the cost per pound at $646 at best, and $2,308 at worst, depending on what was included in the calculation. The next year, 1986, NASA had scheduled the Shuttle for only 14 flights, and the

Challenger explosion in January of that year, of course, forced an adjustment in this schedule.[21] Challenger was only the 25th flight of the shuttle.

Congressman Bob Traxler of Michigan, whose subcommittee oversees independent agencies such as NASA, summarized the issue: "I have a sense that too often on major science projects, not within NASA only, but . . . whether it is DOD or . . . the Department of Energy . . . there is a tendency to low-ball cost estimates in order to get the Congress to buy into the program. Once the program gets launched, it carries with it a number of constituencies who are very supportive of the program, irrespective of its goal and changing role and increasing costs. We find ourselves then over a period of several years looking at an original project that we talked about at X hundred million, not unusual in the case of these super science programs, quadrupling in cost or even more so."[22]

Ernest Fitzgerald had an insightful comment on this sort of thing based on decades of fighting costs on Pentagon boondoggles: "Once you get into these programs they all follow what I call Fitzgerald's First Law of program management. It holds that there are only two phases to a major program. The first phase: 'It's too early to tell.' The second phase: 'It's too late to stop.' You notice that on the bomber programs especially. The B-1 and the B-2. Keep the details and the circumstances secret, their status, everything else. During the secret phase, they always say, 'It's too early to tell, extending the frontiers of science, that sort of thing.' Until all of a sudden they say, 'Hey, we have a few problems, but it's too late to stop.' It's a simple mechanism, and it works most of the time."

The Hubble Space Telescope illustrated Fitzgerald's First Law. In a 1980 report, 10 years before the telescope was launched, the GAO warned: "NASA should provide the Congress with better financial information on the project, including information on program reserves, contract costs, and more accurate development and life-cycle estimates. . . . NASA officials strongly oppose publishing detailed financial data in such quasi-public documents as the project status reports that, in their opinion, would prejudice the Government's negotiating position. They recommend that GAO omit explicit data on reserves from this report." What are these "reserves"? Extra money set aside, in case the project proves more time consuming and difficult than originally estimated.

These reserves weren't petty cash, even in 1980. The base contracts added up to $230 million, but the reserves came to $296 million.[23]

Why doesn't NASA want to tell Congress about the reserves? A NASA Comptroller in 1978 told Congress, "It has been . . . my experience . . . that if one ever identified any contingency as such, that was the first thing that came out of a project." Another reason for keeping the reserves secret, according to the GAO, was that "if the contractors were to find out what the reserves are they might tend to want to spend them." But NASA also wanted to keep the potential costs secret from Congress. A former NASA administrator is quoted in the GAO report: "Current contract target and ceiling prices are subject to readjustment. The contractor's estimated price at completion will change as the scope of work is impacted by development events."[24]

When the program received its first money in fiscal year 1978, the telescope was scheduled for launching in December 1983. However, delays developed. By 1982, Congress had pumped $470 million into the project, and of course, having spent that much money, it was too late to stop. "NASA rescheduled the launch date to the first half of 1985," said the GAO report.[25]

Part of the reason Fitzgerald's law works is that Congress, to create patronage jobs back in home districts, becomes a tacit coconspirator. Naturally, no senator or congressman will admit to compromising his decision making by the desire to enrich his own constituency and, incidentally, perhaps to collect some campaign contributions from the contractors. Senator Albert Gore Jr. explained the public relations gambits at a hearing on the Hubble fiasco: "Now, the relationship between NASA and the Congress is part of the problem, because there are certain games that get played on both sides, and NASA learns from experience that one object of the game is to get the leading wedge into the budget process. Get the new program underway, and then the funding growth will naturally continue.

"Congress then comes back and says, okay, that is fine. We have put out the press releases, we have excited the imagination of the American people, we have encouraged them to dream these wonderful dreams, but now budget reality forces us to spread a smaller amount of money very thinly over all these projects, and that means you have to scrimp and save, cut out the testing if that is needed, to save money on Hubble.

"Cut out this, and cut out that, but do not interfere with the next dream. Let us go ahead and get started on the next big, extravagant dream, and we will get the start-up funds by cannibalizing the dreams that were announced some time ago but are now in the unglamourous development process."[26]

THE OVERSELLING OF STAR WARS

Congress's need for patronage isn't the only force shaping how science projects are funded by the government. The President also pushes science beyond the sobering constraints of peer review at times. Star Wars is a classic example. President Reagan announced the SDI plan in a March 23, 1983, speech: "We could intercept and destroy strategic ballistic missiles before they reached our own soil or that of our allies." He urged the scientific community, "those who gave us nuclear weapons to turn their great talents now to the cause of mankind and world peace: to give us the means of rendering these nuclear weapons impotent and obsolete."[27] This overblown and highly optimistic speech was the result of oversell.

Perhaps nothing subverts public confidence and trust in science more than the gross exaggeration of what the scientific project can and will achieve. Oversell, which seems to be a critical characteristic of nearly every super science project, is destructive to science for two reasons. First, it involves a massive abuse of trust. One or more famous scientists, backed up by hordes of lesser lights, push a position that has little or no scientific justification. Researchers who oppose the oversold projects and speak out for reason and moderation find themselves in the unpleasant and often untenable position of opposing some of the most famous people in the field. In the end, after the project fails to live up to its billing, those who pitch the overblown project undercut public confidence in science and scientists, and those who speak out for reason waste much productive time fighting absurdities. Second, oversell squanders vast sums of money that might go into better science projects, or perhaps shouldn't be spent at all. If the scientist pitchmen can get the congressional and presidential patronage machine behind them, literally billions can be wasted pursuing the impossible.

Although some early designs for the Strategic Defense Initiative were nonnuclear, the most spectacular example of oversell concerned a possible nuclear version, centering on a device known as the nuclear pumped X-ray laser. Two facts make the overselling of the nuclear pumped X-ray laser an exceedingly important case. First, documentation has shown the abuse of trust to be on a scale that goes well beyond that of other super science projects. Second, the amount of money involved is so vast that it dwarfs all other post World War II overselling.

The case began with one of the most famous pioneers of nuclear weapons, Edward Teller—the physicist often referred to as the "Father of the H-Bomb." Teller had extremely high credibility within the Reagan Administration. Secretary of Energy John Herrington said: "My personal feeling as Secretary of Energy is that Dr. Teller is a national asset. His reputation and what he's done for this country and the scientists that were with him after World War II is something we can never forget. Most Americans would feel that way. His reputation speaks for him and his credibility is very, very high. I put my money on Dr. Teller."[28]

So apparently did Ronald Reagan. Teller, along with an associate from the Lawrence Livermore National Laboratory (LLNL), Lowell Wood—often referred to as Teller's protégé—reportedly played key roles in the events leading up to Reagan's speech.

At about the same time that Reagan took office, *Aviation Week & Space Technology* ran an article describing a "breakthrough" at the Lawrence Livermore Laboratory that potentially could "blunt a Soviet nuclear weapons attack on the U.S."[29] The article continued: "In a recent test at the Nevada underground nuclear test site, the laboratory demonstrated—in a vacuum chamber simulating space—a very small, compact laser device pumped by X-rays from a small nuclear detonation." The article went on to describe an X-ray laser that produced radiation beams so powerful that they evaporated the target surface. The breakthrough described was even more miraculous in that the project had cost a mere $10 million. The next several pages of the article described in detail what amounted to the whole Strategic Defense Initiative complete with artist's rendering of one of the weapons poised for action in space.

There was something to what the article said, according to Roy Woodruff, head of Nuclear Weapons Research at the Lawrence

Livermore X-ray laser lab at the time the article was published. The concept of pumping a laser with a nuclear explosion had been demonstrated, in principle. The lasing would take the explosive force of a tiny portion of the otherwise more or less uniform ball of energy emanating from a nuclear explosion, and direct it for a fraction of a second through a laser to amplify its force in a specific area. "In the vacuum of space, this *would* mean, if a weapon could actually be developed, that, hypothetically, an object only tens of miles away from the unfocused side of the blast would not be significantly harmed, but an object one thousand miles away which was the target of the lasing might be severely damaged or destroyed," said Woodruff during an interview in Santa Fe, New Mexico, on February 1, 1991. However, Woodruff emphasized that demonstrating a phenomenon in principle is a very long way from creating a nuclear pumped laser that would work in practice. Continuing the research seemed to be an extremely good idea. Betting the national treasury on practical application for the device was another matter altogether.

Ronald Reagan may not have been a regular reader of *Aviation Week,* but there is no question that Edward Teller made the President aware that something had been achieved. Teller later said, "I don't know how many times I met with the President. But they were very brief meetings. They were insignificant. And they have been blown out of proportion."[30] However, on October 3, 1982, Lawrence Livermore Laboratory scientist Hugh DeWitt wrote David Saxon, President of the University of California that Teller and Wood reportedly had met with the President with the aim of increasing the funding on the X-ray laser program to $200 million annually. DeWitt then referred to an unclassified informal speech Teller had given to LLNL personnel in which he had indirectly referred to the X-ray laser as well as other nuclear devices as "third generation" weapons that potentially had the capability of affording "an effective nuclear shield for the nation."

D. M. Ritson quoted Teller in an article in *Nature,* "What this lab can accomplish now is more important than what we have ever accomplished before. The third-generation effort [on nuclear technologies] gives us every expectation of an effective nuclear defense, and if defense by nuclear weapons is possible, we must have it."[31]

Teller also ballyhooed the X-ray laser program to the public. On October 13, 1982, he wrote in the *Pittsburgh Press:* "The possibility of very effective defensive weapons comes from the prospect of using nuclear explosives of limited energy to disable incoming missiles. This could be accomplished in a manner that would leave the ground level free of the effects of blast, heat and fallout."

Roy Woodruff, on March 4, 1987, wrote the head of the Department of Energy's San Francisco Office, Richard Du Val about the "long tradition" in the United States of "our best scientists" directly discussing with the President the "pivotal" national security implications of new technologies. The classic example of this, Woodruff pointed out, was Einstein's letter to President Roosevelt, which, along with Roosevelt's discussions with other scientists, led to the Manhattan Project. Woodruff acknowledged that he couldn't be an eyewitness to what Teller told Reagan because he, Woodruff, was not present at the meeting. Nonetheless, Woodruff concluded, "I believe it is reasonable to assume that the discussion between Edward Teller and President Reagan was a major factor in the decision to proceed with the Strategic Defense Initiative."

Perhaps the closest to a clear statement of what Teller told Reagan in their meeting appears in portions of a September 25, 1982, letter from Teller to Reagan. These remaining portions of this letter, which was heavily edited by the censors during declassification, make reference to an important meeting only a few days earlier. After thanking President Reagan for his "thoughtful attention" and expressing the hope that he answered the President's major questions, Teller politely indicated that he anticipated eagerly a "favorable" and "timely" decision by the President for "American exploration of 'Third Generation' nuclear weapons technology at a pace commensurate with its promise for replacing" the existing deterrence doctrine of Mutually Assured Destruction with a new doctrine of national defense which Teller described as "assured survival."

Teller did not claim the Strategic Defense Initiative could stop all missiles. He, himself, commented on that minutes after President Reagan's speech announcing Star Wars. "Secretary of State George Schultz, with whom I worked for many years, asked me: Could we stop 99.9 percent of the incoming missiles? . . . The answer is no. . . . But defense does not need to be a tight umbrella."[32]

Teller's remark might lead the reader to infer that perhaps an umbrella with a few leaks in it *was* feasible, and this would obviously be better than no umbrella at all. Teller would later say on Lehrer/MacNeil, on August 5, 1986, "I am talking about protection. I wish to make it as complete as possible. But something like complete protection does not exist—it is postulated by those who want to disprove it, who say all or nothing and all being impossible, therefore nothing."

Reagan's speech apparently was put together by "a small, inner circle of staff, bypassing the normal channels of review that might have pointed out potential problems," wrote one expert on the matter with the collaboration of a key congressional staffer.[33] According to this source, which surveys the other research into the speech, George Schultz, the Secretary of State, along with Richard Perle, Assistant Secretary of Defense, tried to "stop or alter" it. The Joint Chiefs of Staff were appalled that their idea for more research into the area was being puffed up into the forthcoming extravaganza. The President's science adviser, George Keyworth, "a Teller protégé, recommended by Teller for the science adviser job,"[34] apparently did not tell his boss that quite a few weapons-building scientists were highly skeptical about nuclear missile defenses. Since virtually none of the scientists, and few of the political advisers, who should have been advising Reagan on this speech had done so, exactly what did Edward Teller say to the President? We may get a pretty good inkling from the declassified portions of letters he wrote to Reagan's Science Adviser, Dr. George Keyworth, Reagan's National Security Adviser Robert McFarlane, and Reagan's Arm's Negotiator, Ambassador Paul Nitze. These letters ultimately led California to pass a Truth-in-Science law, discussed later in this chapter.

THE TELLER LETTERS

In June 1988 two Democratic congressmen, Ed Markey and George Brown, successfully asked the Department of Energy to declassify the letters, with whatever technical deletions were necessary. The key portions of the first Teller letter, written on December 22, 1983 to Reagan's Science Adviser, George Keyworth, began with a classified "Christmas present." This was, according to the letter, a "quantitative" proof that certain measurements from an X-ray laser nuclear test had been obtained. In fact, Teller wrote that X-ray lasing was the only way the results could be explained. He then added a key phrase, which would echo throughout the subsequent controversy, ". . . we are now entering the engineering phase of X-ray lasers. . . ." He followed this with a request for an immediate $50 million more for the project and an increase for the next year, 1985, of $100 million. The results provided by the nuclear tests didn't mean that the scientists at LLNL are "geniuses," wrote Teller. Rather, "too many people may have overestimated the difficulty of the job." In addition, Teller wrote that the nation faced a "potentially dangerous situation" because of "evidence" the Soviets were "a few" years ahead of the United States on this type of project.

Another worker on the X-ray laser project was Roy Woodruff, the head of Nuclear Weapons Research at the Lawrence Livermore X-ray laser laboratory. "I was given a copy of the Teller letter by a source within the lab, who was so appalled at the content that he hoped I would do something. But the classification rules prevented me from going public. Even the mention of the name of the project, Excalibur, was classified at that time," said Woodruff. "The name 'Excalibur' came from a movie about King Arthur which was playing about that time and which some of the people at the lab went to see," recalled Woodruff in an interview with the author.

Classification rules prevented Woodruff from blowing the whistle, but he could respond with his own classified letter. So six days after Teller wrote George Keyworth, Woodruff also wrote to Reagan's Science Adviser. [The letters were never sent, as we shall see further on.] Teller had referred to "a few others" working on the X-ray laser; on December 28, 1983, Woodruff wrote as the leader of that research. He was able to explain the physics to Keyworth in detail because the Science Adviser's PhD was in physics; before he took the Science Adviser job, Keyworth was the Physics Division Leader at Los Alamos. "I wish to 'set the record straight' and mitigate some of what I perceive to be premature conclusions arrived at by Edward," wrote Woodruff. He objected to a statement deleted in the sanitized version of Teller's letter, which said something about the data being in "essentially quantitative agreement," with computer models developed at Lawrence Livermore.

There is some dispute over the meaning of "quantitative" as opposed to "qualitative" in this context. According to the GAO, "In general quantitative means that the results were 'close' to the predictions, and 'qualitative means the results were 'not as close.'"[35] Woodruff and other LLNL scientists emphatically intended the ordinary meanings of these terms. Specifically, "quantitative" implied that something could be numerically measured, and that is precisely what the X-ray laser researchers could not do.

Woodruff wrote Keyworth that he was "hesitant to claim quantitative agreement at this time." He wrote instead about how "the status can be more accurately stated." Critical characteristics were indeed measured, said Woodruff, and were "in solid *qualitative* agreement with predictions." These characteristics of the output of the nuclear pumped laser were those of spectral temporal, spatial, and intensity. But these qualitative data, although considerable, still didn't give the LLNL scientists "a solid predictive ability based on current models and codes." As a result, Woodruff emphasized, "[M]any physics questions remain." Most important was the practical consequences of this data. Even though there was no question of the nuclear blast "clearly demonstrating lasing," the data in hand "do *not* establish that these systems can be scaled to the range needed for military applications."

Woodruff's letter clearly meant that some X-ray lasing had occurred, but how much could not be determined (i.e., the lab had qualitative but not quantitative measurement). This led to another of Woodruff's major points, "*The X-ray laser is nowhere near the engineering phase at this time* [emphasis added] . . . critical physics characterization and scaling experiments must be carried out before we can attempt to assess the weapon feasibility of this concept. Only then will we possibly be at the beginning of the engineering phase."

Woodruff then made his own request for money, suggesting that with more cash the "milestone of assessing weapon feasibility" could be moved up. He concluded his letter with a strong emphasis on the success achieved so far—nuclear pumped X-ray lasing, that is, a *directed* nuclear explosion had, in fact, been achieved. This caused both the enthusiasm of the LLNL scientists and the "need for accelerating this research . . . to grow." But he warned, "[I]t is premature to extrapolate present successes to the conclusion that a viable weapons system is possible in the near term."

In an interview with me on February 1, 1991, Woodruff elaborated on this: "With unlimited money, we could get to a critical milestone within five years, or within ten years at the rate the program was funded in 1985. What we would have at that point is we might have proven, or definitely not proven the feasibility of Excalibur—an H-bomb one million times brighter at a point in some precise direction." Measured in political terms, the laboratory was at least one and maybe two and a half presidential elections away from the "engineering phase" of the X-ray laser. Reagan might be out of office before the researchers had any idea that this device could work as a weapon in the SDI program.

What could the weapon actually do? "Excalibur is an antisatellite weapon; it makes SDI very difficult," says Woodruff. It is an antisatellite weapon because its range is only a few thousand miles—too short to be effective against attacking missiles themselves during the few minutes available for defense. Since virtually all the plans for Star Wars, at that moment, relied on satellites, the implication of the X-ray laser as an antisatellite weapon was that strategic nuclear defense wouldn't work—the exact opposite of what Reagan had called for in his speech. The X-ray laser could be rocketed up through the atmosphere in a few minutes and could destroy Soviet SDI and other satellites that might be thousands of miles away. If the Soviets had an X-ray laser, it would be similarly used against American satellites, including those for strategic nuclear defense. However, a massively more powerful laser "popped up" straight into space, might in principle, be capable of doing more than zapping satellites. It might be able to take advantage of the curvature of the Earth to shoot down at the actual attacking intercontinental ballistic missiles (ICBMs) as they rose through the atmosphere, even if they were as far away as 10,000 miles from the exploding X-ray laser. "In my opinion, that's why they came up with 'Super Excalibur,'" recalls Woodruff.

On December 28, 1984, one year after writing the letter to Keyworth, Teller wrote about Super Excalibur in two more letters, one to Ambassador Paul Nitze, the U.S. Arms Negotiator, and the other to Robert C. (Bud) McFarlane, Reagan's National Security Adviser.

In his letter to McFarlane, Teller says that the X-ray laser could "locally enhance the brightness and effectiveness of the nuclear bomb effects a million fold." As a result, wrote Teller "it has become highly probable" that it could "destroy sharply defined objects" as far away as "1000 miles and possibly more." The letter to Nitze stated that the device could "kill a target" as far away as 10,000 km, or 6,200 miles. The same target, Teller explained, would not be destroyed directly by the bomb itself, without the

X-ray laser device, unless the bomb were only 10 km away. He put the change into perspective when he wrote that the enhancement of the explosive force, in a given direction, is "thus comparable in magnitude to that involved in moving from chemical to nuclear explosives." Whether or not the forward movement is as great as Teller says, it certainly is an advance that seems to increase from letter to letter. In the same letter, he described the X-ray laser as a trillion times more powerful than the hydrogen bomb itself, in a given direction pinpointed by the laser. According to Woodruff, a million times enhancement was the intention of Excalibur, a trillion times was the dream of Super Excalibur. The letters don't say what numbers Teller may have discussed with Reagan.

Although both Excalibur and Super Excalibur were fictional weapons, Super Excalibur did differ from Excalibur in two respects. First, Super Excalibur would be much more powerful. Second, although Excalibur existed as a positive, although so far unmeasurable, test result and as a research project with well-defined milestones, it did not exist as a weapon. By contrast, Super Excalibur did not exist in any way at all, except as a name and a concept.

Woodruff wrote to Deborah Grimes, of Congressman Pete Stark's office on October 3, 1988, concerning Teller's assertion that "theoretical calculations indicate" tremendous benefits from Super Excalibur. Woodruff pointed out that persons with a fair knowledge of the design of weapons would take Teller's claim to mean that a "design concept has been computer simulated." For this to be the case, explained Woodruff, the researchers would have to have three things—"sufficient definition of the concept," "adequate computer models," and adequate "computational capability." Otherwise they couldn't do the analysis for the design of the device. According to Woodruff's 1988 letter to the congressman, the LLNL did not have this knowledge in 1984, "nor even today." Although the "overall concept" had been sketched out, the requisite calculations could not be done because the underlying "details . . . were *not* understood." Woodruff concluded, "Super Excalibur was *so* conceptual, that it could not be run through a computer simulation."

Teller told McFarlane that Super Excalibur, with moderate support and "considerable luck . . . might be accomplished in principle in as little time as three years." Woodruff, the head of Nuclear Weapons Research on the project, had estimated 5 to 15 years just to establish feasibility for the more modest Excalibur, and another 5 to 10 years beyond that for engineering development.

Teller wrote to Paul Nitze about Super Excalibur, "[A] single X-ray laser module the size of an executive desk which applied this technology could potentially shoot down the entire Soviet land-based missile force, if it were to be launched into the module's field-of-view."

Woodruff again wrote a letter to Nitze to set the facts on the record in January of 1985. He explained his reasons for doing so in his October 3, 1988, letter to Congressman Stark's office. Woodruff was "concerned" that Nitze would be "further misled" by Teller's claim concerning theoretical calculations for Super Excalibur. Woodruff found the context of Teller's reference to an executive-desk-sized version of the weapon as relatively close at hand to be "especially" misleading, since the requisite calculations for this claim "did not exist." However, Woodruff wrote the congressman, "the claim that they did provided substance to Teller's arguments that the negotiators should base national arms control policy on this eventuality."

Woodruff wrote in his January 31, 1985, letter to Nitze that he was concerned that the balance is "overly optimistic" in Teller's letter and in possible "additional discussions" with Lowell Wood. He then summarized the results of the X-ray laser experiments and said that they "make clear a number of points." Among these points is that the physics, the computer codes, and the data base that they had at that time "are only capable of guiding our endeavors in a qualitative manner." The LLNL researchers would have to experience "large advances" before they could successfully make "quantitative predictions." Key experiments involving the physics and the scaling of the device would have to take place before the LLNL scientists "can fully assess the weaponization potential of the X-ray laser concept." Woodruff stated that he himself "fully" expected the key experiments would "establish that the X-ray laser could be an effective weapon." However, until the experimental results actually panned out, "*the issue remains a matter of speculation* [emphasis added]."

As for the feasibility of an executive-desk-sized Super Excalibur, Woodruff wrote Nitze that Teller's opinions "may be interpreted with too much optimism." Although Teller's executive desk statement was "technically correct insofar as the realm of possibility," the Teller statement "does not convey the difficulty of such a weapon achievement." Could the LLNL weapons researchers ever provide something even "close" to what the executive desk statement suggested? "Not impossible, but very unlikely," wrote Woodruff.

In the interview on February 1, 1991, Woodruff elaborated, "Super Excalibur didn't even exist as a program goal because we didn't know if it was scientifically feasible. The most you could say about it is that it didn't violate the laws of physics."

Woodruff had a clear enthusiasm for the X-ray laser project, but he also clearly understood how difficult it would be to realize the device. He wasn't the only one. The GAO reported in 1988, "Dr. George Miller, the current LLNL Associate Director for Defense Systems, supported Mr. Woodruffs views and stated that the X-ray laser was not ready for engineering then or now."[36]

In fact, knowledgeable outsiders were publishing many of the same cautions that Woodruff expressed in classified meetings. In a 1985 article in *Foreign Affairs,* former Secretary of Defense Harold Brown wrote of the X-ray laser: "Proof of the most basic principle has been established, in that bomb-driven X-ray lasing has been demonstrated to be possible. But there is doubt as to what intensity has been achieved; it is in any event far less than necessary for use in active discrimination, let alone target kill. Demonstration of the physics of a possible weapon is at least five (more likely ten) years off. Weaponization would involve another five or more years, and only thereafter could its incorporation into a full-scale engineering development of a defensive system begin."[37]

Exaggerated claims are common ploys as part of the foot-in-the-door buy-in for super science projects. As noted earlier, the Space Shuttle was oversold, as was the Hubble Telescope. In 1983 the Department of Energy actually abandoned an earlier version of the Super Collider, ISABELLE, after an expenditure of $200 million and completion of 75% of the conventional construction. The big magnets for it simply could not be made to work.[38] The people who routinely oversell Congress probably do believe that, with a great deal of luck, they can deliver more or less as promised. When they fail, they nearly always give the same excuse. "Let me plead guilty to the great crime of optimism," said Teller some years after he wrote his letters.[39]

However, many researchers who were deeply immersed in the X-ray laser project believed Teller's claims went far beyond the usual garden-variety optimism routinely found in super science buy-ins. One such scientist was George Maenchen, a strong proponent of the X-ray laser, who had devoted his working life for the previous 10 years to technical aspects of the device. During that time he had been the senior designer and for part of the time was also the group leader for designing the X-ray laser. Describing the overselling claims for the X-ray laser as "outrageous" and "absurd," Maenchen wrote in a December 23, 1987, letter to James Ohl, a GAO investigator, that the claims were in the sphere of vivid imagination. They didn't merely show someone's own excitement for the project and tendency to dwell on the most hopeful aspects of the situation. Maenchen's letter, it must be noted, emphasized his strong commitment to the X-ray laser program as a valuable research project with a certain amount of, but far away, potential for military application.

In a September 7, 1988, memo to LLNL Director John Nuckolls, Maenchen elaborated on the distinction between optimism and the degree of exaggeration in the Teller letters, pointing out that strong advocates might magnify advantages and not fully discuss expenses or technical problems. Maenchen wrote that government officials most likely routinely discount that usual sort of optimism. In this case, however, the extent of the magnification presents a difference, as does the fact that the listeners may be unfamiliar with the science and technology. For example, Maenchen wrote that if someone claimed he would give an engine a million times increase in horsepower, the listeners might be exceedingly skeptical. But most people are not familiar with X-ray lasers and what the evidence shows they can do. So, when a world-famous scientist endorses or makes assertions that greatly depart from the knowledge, understanding, and expectations held only by those with highly specific scientific expertise, the less knowledgeable listener may not give the claims careful and exact evaluation and may instead take them at their apparent significance. As with the earlier letter to the GAO, Maenchen again closed by stating his strong support for the X-ray laser program, which he described as basically a research program, but one that needed to be energetically pursued to prevent the United States from being caught unaware by a new technology in the hands of its enemies.

By 1987, even Teller was making less optimistic statements about the X-ray laser. He was quoted as saying, "Whether or not I said the X-ray laser was in engineering phase is not relevant, and I don't want to waste time on this subject any more." Teller added,

"I say that we will know about the costs and if it will work as soon as we deploy."[40] His emphasis on deployment is critical to an understanding of exactly how he was promoting the X-ray laser.

In his classified, December 22, 1983, letter to Presidential Science Advisor Keyworth, Teller had written: I do not believe that the X-ray laser is clearly the only means, the best means, or even the most urgent means for defense. It is clear, however, that it is in this field that the first clear-cut scientific breakthrough has occurred. It is necessary to draw all possible consequences from this fact. . . ." Thus, Teller's commitment was clearly to the notion of nuclear defense, that is, Star Wars, itself, rather than to one particular device, the X-ray laser. With the launching of a multibillion-dollar program for Star Wars based on the X-ray laser, the underlying concept of nuclear defense would be in Fitzgerald's "too late to stop" phase of the program. After the truth could no longer be ignored about how far away the X-ray laser was from possible use, the SDI program switched to a new centerpiece, Brilliant Pebbles, discussed in Chapter 7.

WOODRUFF'S DISSENT

Even though the truth finally prevailed over Teller's gross exaggeration about the X-ray laser, Woodruff's letters had absolutely no impact because, in fact, the letters were never sent. According to the 1988 GAO Report, the then Director of Lawrence Livermore, Dr. Roger Batzel, asked Woodruff not to send them. Concerning the letter to Keyworth, the GAO Report said Batzel felt the letter wasn't necessary: "Dr. Batzel told us that Dr. Keyworth is a knowledgeable physicist and had been briefed on the X-ray laser program. Furthermore, Dr. Batzel said he believed Dr. Keyworth understood that Dr. Teller is a theoretical physicist and, like others, knew Dr. Teller to be a 'technical optimist.'"[41]

Furthermore, the GAO Report, based on comments Woodruff made before writing his letters, claimed that he didn't really disagree all that much with Teller. According to the GAO Report, Woodruff had previously said some of the same sorts of things. Woodruff was outraged by this assertion.

The dispute gets particularly obscure because it concerns classified documents and meetings, so only a few elements of it can be discussed here. A key piece of the GAO's case concerns a January

13, 1984, memo sent by Woodruff to a number of Energy and Defense Department officials as well as to Keyworth. According to the GAO, in the memo Woodruff used the phrase "excellent quantitative agreement with predictions" for two of the factors Teller had also cited as in "essentially quantitative agreement." "In this memorandum," said the GAO Report, "Mr. Woodruff made statements similar to some of Dr. Teller's statements, which Mr. Woodruff had questioned." This is "a real distortion," according to Pete Didisheim, aide to Congressman George Brown, who requested the GAO Report in the first place. "The GAO position is simply unsupportable," wrote Woodruff to Congressman Pete Stark's office. During the Santa Fe interview in 1991, Woodruff explained: "Although I did use the term 'excellent quantitative agreement with predictions' as the GAO reported, I significantly restricted its application. When you measure the performance of a laser you measure color, time-width of the laser impulse, and the intensity. What I was referring to was that we had excellent quantitative agreement in color, but I said we had qualitative agreement on intensity. Teller said we had quantitative agreement on intensity. He was wrong, I was right. What the GAO did was quote me when I was in agreement and not quote my statements when I disagreed with Teller. Furthermore, the GAO never discussed the January 13, 1984 memo with me."

An examination of the subsequently unclassified portions of the key January 13, 1984, memo supports Woodruff's assertions and Didisheim's remark that the GAO report gave a "real distortion." Woodruff actually wrote, "The data from the length study where we measured the output from a . . . laser are in solid qualitative agreement with predictions and may ultimately be in sufficient quantitative agreement" to give a preliminary understanding of how much gain in force the lasers provide. Woodruff's memo discussed the military implications, which it plainly stated "are not totally clear at this time." The memo noted, however, that the device might have use as an antisatellite weapon, but even this use would not be known until after key physics and scaling experiments were conducted. Without these as yet unconducted experiments, the military application continues to be "a matter of speculation."

In the October 3, 1988, letter prepared for Congressman Stark, Woodruff added that the GAO had "unilaterally decided that there is no significant difference between 'quantitative and qualitative.'" This "redefinition of reality by a GAO investigator," wrote Woodruff, resulted in the GAO's deducing that he said the same thing as Teller.

Woodruff did ultimately meet personally with Keyworth on February 15, 1984. "Mr. Woodruff told us that, at this brief meeting, he presented the X-ray laser program's status and clarified Dr. Teller's letter," said the GAO Report. Woodruff, in an extensive 1991 interview, says of the meeting, "I had access to Keyworth for ten or fifteen minutes and tried to explain the current status of the program. But after that Keyworth gave a speech on the X-ray laser, and his speech was inconsistent with my position."

It was much more consistent with the contents of the Teller letters. In December 1985, Keyworth addressed the "Lasers '85" conference in Las Vegas. The LLNL newspaper, *Weekly Bulletin,* reported portions of his speech: "'The President's commitment of SDI is rooted in the rapid and significant advances in the field,' Keyworth said, pointing out that two years ago scientists in the field were asking if we could develop an effective SDI system. 'Now we're asking how best to do it.' . . . A single X-ray laser could defend against the U.S.S.R.'s entire offensive forces. . . . '"[42]

Batzel also did not want Woodruff to send his letter to Nitze, preferring that Woodruff make his points in person. His concern, according to the GAO Report, was with Woodruff's "making budgetary comments and requesting specific funding" in the letter. The GAO Report fails to explain why Batzel didn't simply ask Woodruff to remove the parts of the letters dealing with money.

Woodruff, in the Santa Fe interview, said: "The GAO's claim of what Batzel said doesn't even pass the "Ho-Ho" test. . . . One cannot describe times for achieving program goals without simultaneously stating funding requirements. You have to link funding rates with milestones. For Batzel to say I shouldn't write about money is ludicrous. Besides which, he never said it to me, which is why he simply didn't edit it out of my letters.

"The reason Batzel gave me [for not sending the letters] was that he didn't want the controversy I had with Teller going outside the laboratory. He didn't elaborate on this beyond not wanting my views in writing. What I wanted was, since Teller's letters were on the table, to put my letters on the table as well. It would do in the same file and if you got one, you'd get the other. And that's the only thing Batzel didn't want me to do."

Here's how the GAO reported Batzel's reasons for doing what he did: "According to Dr. Batzel, there was nothing in Dr. Teller's letters which violated the laws of physics." Also, Teller had discussed some of the ideas as "in principle," and thrown in "many qualifiers."

In his October 3, 1988 letter to Congressman Stark, Woodruff wrote that although the laws of physics weren't violated by Teller's assertions, this fact "does not mean that a workable weapon can be achieved—and certainly not in the near term."

Woodruff did meet Nitze on February 7, 1985, for a couple of hours, and was able to discuss Dr. Teller's letter "in considerable detail and had ample opportunity to state his views," said the GAO report. But this meeting did not leave a paper trail—no letter that might dog those of Edward Teller and perhaps find its way to congressional critics of the program. Key congressmen who opposed SDI did not get the decisive information that they needed at the time to squelch the whole Star Wars concept.

Yet the critics were there, waiting. Responding to a *Los Angeles Times* story of November 12, 1985, about a possible design flaw in the instruments used to measure the X-ray laser nuclear blasts, Congressmen Bill Green and Ed Markey asked the GAO to investigate "the management of and funding for the X-ray laser program."

The June 1986 GAO Report that followed said, "Essentially, we found the X-ray laser program is a research program with many unresolved issues. In our opinion, there was no 'design flaw' in the diagnostic instrumentation. . . .[43]

"What happened," said Woodruff in Santa Fe, "is that the instruments got hot which made their measurements subject to error." Obviously, the original story to the *Los Angeles Times* had been leaked by critics of the program. Their leak apparently was to counter previous leaks by supporters of the program, including Teller. Teller told a student audience in California on April 3, 1985, that X-ray lasers "exist not on paper," but in reality. "Three weeks ago, I couldn't have told you that," said Teller.[44] Slightly more than one month later, the Lawrence Livermore Laboratory declassified the statement. "The scientific feasibility of the nuclear explosive driven X-ray laser has been demonstrated through testing conducted at the Nevada Test Site."[45] So Teller had apparently leaked this information a month earlier.

Woodruff had other problems with Teller besides the latter's extraordinarily optimistic letters. One of them is detailed in a *Los Angeles Times* story by Robert Scheer, concerning a September 6, 1985, meeting in the office of Lt. General James Abrahamson in the Pentagon. "Scheer describes it accurately, no one ever disputed any part of that account," said Woodruff in Santa Fe. The Department of Defense was about to transfer $100 million from its budget to the Department of Energy to work on directed nuclear weapons such as the X-ray laser. The money was to be divided up among three laboratories—Lawrence Livermore, Los Alamos, and Sandia. Top officials and key weapons lab scientists such as Woodruff and George Miller, from Los Alamos, had agreed on a three-way split: $60 million for Livermore, $30 million for Los Alamos, and $10 million for Sandia. "Everyone was in agreement," Woodruff said, "[Abrahamson] raises his pen to the board to write the numbers, and in walks Dr. Teller." According to the Scheer article, as confirmed by Woodruff, "Teller said he wanted the entire $100 million to go to Livermore. The team at Livermore was 'very bright and enthusiastic,' Teller argued. And more important, President Reagan had promised all the money to him. . . . Without further discussion, Abrahamson announced that the entire $100 million in new funding would go to Livermore and that the other labs would be expected to get their money from other, unnamed sources. Woodruff, who was sitting on a couch with two representatives from Los Alamos, says he sank into his seat with embarrassment at Teller's intervention."[46]

According to Woodruff, the story demonstrates how Teller was undercutting his authority. "It shows the person in charge of the X-ray lab wasn't really in charge," recalled Woodruff. The story also shows how political funding for the program was beyond anything that could be defined as "peer" reviewed.

Department of Energy Secretary Harrington discounted the story. "One of the big allegations was at a certain meeting, one hundred million dollars was divided 100/0 instead of 60/40. That's nothing in a multi-billion dollar program."[47]

Embarrassed at key meetings and unable to make a documented rebuttal to Teller's letters, Woodruff resigned in protest as Associate Director for Defense Systems. In an originally classified October 19, 1985, letter to Lab Director Roger Batzel, Woodruff raised the question of whether he should keep quiet, in which case the grossly exaggerated statements on the X-ray laser discussed earlier could "continue" to "potentially mislead' "the highest levels of leadership" in the United States. Woodruff stated that keeping quiet would certainly be the easy way to do things "[A]s you have made it clear on numerous occasions that your preference is for me to do nothing." However, he had to act because "It is simply unacceptable to allow the continued selling of the X-ray laser program . . . And believe me, the selling has not stopped. . . . there seems to be no let up in sight."

On April 3, 1987, Woodruff filed a grievance with the University of California against Batzel. In his grievance statement addressed to UC President Gardner, Woodruff made a number of points. Among them was that Teller had "undercut" his, Woodruff's, management responsibilities on the X-ray laser and had "conveyed both orally and in writing overly optimistic, technically incorrect statements" concerning the X-ray laser to top federal officials. Woodruff said that Batzel was "fully aware" of Teller's actions, and that Batzel had "refused to transmit correcting technical information or allow me to do so" for several of Teller's letters.[48]

His grievance was rejected, but he filed a second one with the Lawrence Livermore Laboratory on April 30, 1987. This time he won.

THE DISPUTED GAO REPORT

The GAO in a report on the controversy stated, "We asked selected LLNL scientists, who had specific knowledge about the X-ray laser program, for their opinions as to the accuracy of the statements challenged by Mr. Woodruff. From these interviews, we concluded there was no general agreement among these scientists regarding the accuracy of the statements."[49]

This conclusion seemed to anger at least one scientist, George Maenchen, the strong proponent of the X-ray laser program whom I discussed earlier regarding Teller's claim to be "optimistic." Maenchen may have been particularly angered because he had specified his own exact views in a letter to one of the GAO investigators, James Ohl, on December 23, 1987. In that letter, Maenchen referred to an earlier interview Ohl had conducted with him, and reiterated what he had said to Ohl. Specifically, Maenchen wrote that "all" of Teller's assertions that had become a matter of controversy were "totally false." Maenchen went on to write that Woodruff and another LLNL scientist, George Miller, had presented correct accounts of the X-ray laser program, but that different persons had not done so. Among those Maenchen specifically named was Edward Teller.

In a January 14, 1988, letter to Ohl, Maenchen also wrote that the GAO had not been conducting an accurate investigation because those interviewed were for the most part managers. By contrast, so far as Maenchen knew, Ohl had restricted to two persons his interviews with scientists exclusively devoted to the technical details of the X-ray laser, who were not also devoting a noticeable portion of their time to promoting the project. Maenchen repeated his request that Ohl discuss the X-ray laser with others who knew the nonimaginary, existent, and identifiable aspects of the program.

After the report came out, and the controversy erupted over what the laser scientists had said, Maenchen wrote Congressman Pete Stark about his earlier letters to Ohl, which had been acquired by the GAO much earlier than the date of their disputed report and which clearly set out his position. Based on these letters, Maenchen wrote, on April 20, 1989, that it was "absurd" for the GAO to claim he agreed with any of the controversial statements of Teller. Maenchen's letter to Stark discussed a GAO diagram that listed various LLNL scientists and purported to show whether or not they agreed with the disputed Teller statements. He discussed one LLNL scientist, Paul Wheeler, listed on the diagram. According to the GAO diagram, Wheeler agreed in two categories and simply wasn't asked about three others. Maenchen wrote that he had discussed the matter with Wheeler, who told him that although he was not officially asked about the three categories, he himself brought them up and dissented from "all" the disputed Teller assertions. The GAO then paid no attention to his voluntary expressions of dissent and wrote that they hadn't solicited his views on these matters. Maenchen then totaled all the purported responses in the diagram and found over half could not be accepted as true. Without exception, Maenchen found that the unbelievable reports in the diagram supported Teller's disputed statements. Maenchen wrote that this was "unacceptable" and stated his suspicion that it was not the result of inadvertent error but of "systematic bias." As in his earlier letters to the GAO and Nuckolls, Maenchen emphasized in his letter to Stark the need for continuing work on the X-ray laser. It was, he said, a research program that was both valid in its own right and critically needed to keep foreign enemies from gaining a military advantage based on the application of new scientific developments.

The leader of the design group for the X-ray laser, Steve Younger, was quoted in the local paper in the Livermore area, the *Valley Times* of February 24, 1989, as saying, "What the GAO chose to do was take a (complex) issue and reduce it to simple statements. I did not agree with their interpretations of my opinions."

A number of laser scientists were exceedingly unhappy over the time frame for development of Super Excalibur in the GAO Report: "We asked selected LLNL scientists, who had specific knowledge about the X-ray laser program, for their opinions as to the accuracy of Dr. Teller's statement about how soon the Super-Excalibur brightness goal would be achieved. Most of the scientists who offered an opinion regarding the accuracy of the statements felt that achieving Super-Excalibur [within a few years] was conceivable or not impossible, especially if considerable support were available."

One LLNL scientist, George Miller, who was in charge of Lawrence Livermore's X-ray laser program after Woodruff, also never said Super Excalibur could be achieved in three years. He wrote Pete Stark on December 5, 1988, saying he was "concerned that my views as expressed in the GAO reports and briefings may not be well understood. In particular, I do not believe it is possible to reach 'Super Excalibur' brightness goal in 3 years. In fact, I have refused to place a timescale on that development because we do not presently have enough knowledge to accurately predict the timescale. I have consistently stated that it will take at least 5 years and the integral expenditure of at least one billion dollars before we can actually demonstrate whether a much more modest goal is possible. . . . we have not yet measured the theoretically predicted weapon level energy performance." Stephen Younger wrote Stark on November 22, 1988, that he "now" did not believe Super Excalibur could be produced within the three-year time frame given in the disputed Teller statement, and he didn't remember "ever" believing that Super Excalibur would be achieved within that time frame.

The GAO Report ended with the following: "We discussed the results of our review with Mr. Woodruff; LLNL officials (including Dr. Roger Batzel, Dr. Edward Teller, Dr. Lowell Wood, and the new LLNL Director, Dr. John Nuckolls); and DOE officials. They all generally concurred with the information in this report." Woodruff was quite strong in his reaction to this GAO claim. "Charles Bowsher [head of the GAO] has put his signature to a statement that is a lie," Woodruff told Keith Rogers, a journalist who had covered the Lawrence Livermore Lab for more than 12 years for the *Valley Times,* and reported Woodruff's remark on April 18, 1989.

Congressman Stark himself wrote Charles Bowsher, head of the GAO. "I ask that you recall and strike" the highly questioned GAO report to which the battery of top scientists had objected. "The report," wrote Stark, "is inaccurate, misleading and biased." But Bowsher strongly defended the GAO's work in his reply to the congressman. Bowsher wrote, "Your letter also stated that we misrepresented the views of some of the scientists we had interviewed. Our files include a tape recording of one of the interviews and written memorandums for each of the other interviews. Each of these memorandums was reviewed and agreed to by the person involved to ensure their accuracy. While it could well be that some of these individuals have now changed their opinions, we have reexamined our supporting files and have verified that we accurately reported what we were told at the time."

According to budget information supplied by the SDIO Office of External Affairs, when the GAO Report was released in 1988, SDI was funded for $3.6 billion, up from $3.3 billion the previous year, and way up from $1.4 billion in 1985, when Lowell Wood was hand delivering Edward Teller's letters. Between 1985 and July 1990, when yet another GAO Report was issued recommending *against* making a deployment decision until after the next presidential election, SDI had gobbled up $19.8 billion in research money. Of course, these figures were small change compared with the money that would be spent in deployment. A "Phase One" system, which would be at best 30% effective, was estimated to require at least $250 billion.

One of the three GAO investigators who worked on the earlier, controversial report, it was disclosed, applied for a job at the Lawrence Livermore facility only one month after the GAO issued its disputed Report. "How could this not be a conflict of interest?" asked Roy Woodruff. Bill Vaughn, a top aide to Congressman Pete Stark agreed: "It's a conflict of interest to be considering a job with someone whom you're in the process of investigating. It clouds the whole report."[50] A major government report, prepared for Republican Senator Roth and issued in 1987, echoed these comments, stating this sort of job switching "may give the appearance of . . . [the revolving doormen] not having acted in the best interests of the government because they viewed a defense contractor as a potential employer."[51] The report was issued by the GAO, in criticism of the Department of Defense.

Concerning its own case of a revolving doorman, the GAO put out a press release on April 25, 1989, that stated: "GAO's review shows that [the person's] substantive involvement in the work leading to the report on the X-ray laser program was completed in April, more than two months prior to his application for employment at Lawrence Livermore. . . . GAO has reviewed all work to which [the person] was assigned back to his involvement in the X-ray laser audit and through the date he left GAO. That review shows that in no case did [the person's] decision to seek new employment affect the outcome of GAO audits and reports, including the X-ray laser report."

The truth of Woodruff's assertions on how far away from practical use the X-ray laser was has since been made clear to nearly everyone, and the device is no longer a key project in SDI. The Federal Government did authorize a special building for X-ray laser research at Livermore, at a cost of over $62.5 million, but by the time of its expected completion in 1992, there probably will not be enough researchers on the project to fill up the facility. The X-ray laser and a related project had received funding of $210 million when Lowell Wood was delivering the Teller letters in 1985. By 1990, these projects were only getting $187 million.[52] The director of the Office of Weapons Research, Development, and Testing at the Department of Energy said, "You're talking about several hundred people gone from the program, never to return. . . . Since they're very much in demand anyway, the scientists don't come back, because scientists don't like to come back to unstable jobs."[53]

In 1987, Herbert York, a former director of the Lawrence Livermore Lab said, "If I was a lab director now, I would be figuring out a way to better control Teller and Wood. They've gone too far. But it's hard. The President calls Teller to the White House to ask his opinion. We can't tell him what to say. I don't know if anyone in a university context can control this. Maybe, if Boeing or Lockheed was running the program, they'd just be fired."[54]

THE FIRST TRUTH-IN-SCIENCE LAW

By the late 1980s, the overselling of the X-ray laser and Star Wars had reached such an extreme level that the State of California passed legislative language specifically intended to ensure that scientists tell the truth to government officials—the first such instance in U.S. history. Passed in 1988 and signed by the Governor of California, the language was directed at world-famous scientists who work at two of the country's biggest science labs, the Lawrence Livermore Radiation Laboratory in California and the Los Alamos National Laboratory in New Mexico. (Although owned by the federal government, these laboratories have always been administered by the University of California under repeatedly renewed five-year contracts.)

Because of the way the California state constitution set up the University of California system, the new legislative requirements, known as "supplemental budget language" do not have the force of law. However, the state legislature controls the university's budget and can deny funds. Accordingly, the concerns of the legislature in passing this language "must be addressed," said the president of the university in a July 6, 1988 memo.

The legislature required that "research produced by the labs is technically sound and that its meaning not be misrepresented to government officials . . . [and] that the existence of dissenting views within the scientific community be acknowledged and made known to U.S. government officials."[55] The legislators also required the university to set up oversight officials, to be appointed by the president of the university and approved by the regents, to ensure that the spirit of the language is faithfully followed.

But even were it to be successful, the legislation would not curb the pork barreling, revolving doors, and lack of contractor accountability that have made so many super science projects little more than public relations gimmicks and a big waste of money.

CHAPTER 4

DOCTORING DATA: MALFEASANCE IN BASIC RESEARCH

This chapter focuses on the response of the American scientific community to charges of misconduct in science research. The chapter concentrates on the experiences of three scientists—Stephen Breuning, John Darsee, and David Baltimore—that illustrate the scientific community's reaction to such charges. By far the most significant of these is the Baltimore case.

THE BREUNING CASE

In September 1988, Stephen Breuning, PhD, formerly of the Western Psychiatric Institute of the University of Pittsburgh Medical School, faced a federal judge and pleaded guilty to two charges of filing fake research reports on projects paid for by federal money.[1] Virtually all his research reports were bogus—nearly all the experiments they purported to report had not been conducted; there was no basis for believing that what they "reported" was true. These specific research reports, moreover, had provoked frightening consequences. Not only had the reports been relied on nationwide in determining drug therapy for institutionalized, severely mentally retarded children, they had advocated the exact opposite treatment from that suggested by bona fide research. Although subsequent scientists recommended the use of tranquilizers for severely retarded children, Breuning's fraudulent claims advocated stimulants. According to the whistle-blower who brought him to justice, ". . . their behavior becomes worse [with stimulants] and these people . . . [have] very, very difficult behavior problems, sometimes severely self-injuring and consequently [stimulants are] dangerous."[2] Breuning had a major impact on psychopharmacology in this field, contributing 24 out of the 70 papers published on the subject between 1979 and 1983.

The whistle-blower in the case was Breuning's mentor, Dr. Robert Sprague, a professor at the Institute for Research on Human Development, University of Illinois, Urbana-Champaign. Sprague, the Principal Investigator on the grants, testified to three congressional committees, documenting the astonishing ordeal he had been put through for revealing the truth about Breuning to the National Institute of Mental Health (NIMH), a funding organization within the Department of Health and Human Services.

Sprague's Investigation of Breuning

What first aroused Sprague's suspicions? While visiting Breuning at the University of Pittsburgh, Sprague discovered that the researcher was claiming 100% agreement from observations of two nurses judging the movements of institutionalized, mentally retarded children. "It is not possible for two judges to agree perfectly when making such judgments," testified Sprague.[3]

Sprague then obtained evidence of fraud on another matter. Breuning had been a researcher at Coldwater Regional Center, Coldwater, Michigan, and had left that institution in January 1981. He produced a study reporting research supposedly conducted at Coldwater during the two years after he had departed from the Center and was at Pittsburgh. But he had left no research assistants at Coldwater and, as best as Sprague could determine, never returned to Coldwater during the two-year period. Sprague confirmed with another researcher that the study could not have taken place without the other researcher knowing about it and "he knew nothing of such a study."[4] The abstract for the study described 180 evaluations of patients. Sprague demanded the backup data from Breuning, which should have involved 180 data sheets. Breuning could only produce 24, "and even some of these data are suspect," testified Sprague.[5]

Sprague then added up the days of research reported in Breuning's annual Progress Report for work he conducted at the University of Pittsburgh, and the total exceeded the number of working days in the calendar year.

Sprague wrote a December 20, 1983, letter to his NIMH program officer, Natalie Reatig, calling for an investigation, not only of the study described by the abstract, but of the rest of Breuning's work, including that done at the University of Pittsburgh. Sprague mentioned another specific Breuning study where a coauthor had asked Breuning for the supporting data but had been told that "none of the raw data is available."[6] He added in his letter that the question naturally comes up concerning the degree of actual data underlying the rest of Breuning's studies. His letter noted that 30 other coauthors from 9 institutions had produced 57 papers, chapters in books, or entire books with Breuning since 1975. As far as Sprague knew, not one of the 30 coauthors, until Sprague, had gone to the appropriate authorities with both a report of suspicion and evidence for it.[7]

The University of Pittsburgh's Investigation

A month later, on January 17, 1984, an NIMH official wrote Dr. Thomas Detre, Associate Senior Vice Chancellor at the University of Pittsburgh, and requested an investigation. The letter specifically discussed Sprague's allegations about the number of working days while Breuning was exclusively at Pittsburgh.[8]

Nonetheless, the University of Pittsburgh only investigated the fake abstract done at Coldwater, Michigan. The Adler committee, a preliminary investigating committee wrote to Donald Leon, MD, dean of the Medical School on February 17, 1984, "Dr. Breuning admitted to us that statements in the abstract were false. . . . We did not investigate any of Dr. Breuning's work done in Pittsburgh."[9]

Breuning resigned from the University of Pittsburgh and took a job as Chief of Psychological Services at a state institution for the mentally retarded in Polk Center, Pennsylvania. Lorraine B. Torres, an NIMH official, wrote Donald Leon, MD, of the

University of Pittsburgh, on May 8, 1984, "We believe that the report of the Research Hearing Board should contain a review of all of Dr. Breuning's federally-supported research activities at the University of Pittsburgh."[10]

The University of Pittsburgh, however, only looked at whether the Coldwater research in any way influenced any research Breuning did at Pittsburgh. The dean of the Medical School, Donald Leon, wrote to Lorraine Torres, of the NIMH, on July 6, 1984, that the University's Research hearing Board came up with "no serious fault" with any of Breuning's actions at the University of Pittsburgh. So, wrote the dean, he had "no grounds" to begin any formal proceedings against Breuning for anything Breuning had done while a faculty member at Pittsburgh.[11]

Investigation by the National Institute of Mental Health

The NIMH decided to launch its own investigation. They appointed an investigator who, with an assistant, spent two weeks investigating *Robert Sprague.* "It seems that this is a little like shooting the messenger," Congressman Dingell later said. With the investigation focused on Sprague, and mentally retarded children continuing to receive medication based on bogus research, Sprague wrote the NIMH investigator "outlining some of my serious concerns." Among his concerns were "the families of patients involved with the stimulant medication studies." He never received a reply.[12]

Eventually the NIMH appointed a panel that began holding meetings on April 19, 1985, sixteen months after Sprague's initial letter. Sprague testified: "Meanwhile, during the glacier-like pace of NIMH investigation, Breuning was out consulting on the topic of 'Medication for Behavior Control' with mentally retarded people. Twenty eight months after the investigations started Breuning spoke to a Behavior management Committee Conference in Lansing, Michigan, on the topics of 'Medication for Behavior Management' and 'Maladaptive Behavior.' Behavior Management Committees are perhaps as influential as any group of people in determining the treatments for institutionalized people since these committees review and decide about appropriate treatment for very difficult-to-manage residents."[13]

During this time *Science* magazine interviewed Sprague, in February 1986. "I saw the article a few days later and it was good," recalled Sprague. "But *Science* sat on it for 10 months, until Dan Greenberg of *Science & Government Report* phoned them and threatened to publish the story himself. The next day, *Science* phoned me, got an update." *Science* published the story on December 19, 1986. "Twenty-three days later, over Christmas and New Years, they [the NIMH panel] issued their first draft report," added Sprague, "that's not the usual season for any organization which had been moving at a glacial pace suddenly to rush out a report."[14]

The panel finally concluded, according to the Weiss Subcommittee Report on Scientific Misconduct "that virtually all Breuning's work was fabricated and that Sprague's work and accusations were beyond reproach."[15]

The University of Pittsburgh investigated Breuning again—after the NIMH began its investigation. According to the Weiss subcommittee report, the new investigation "included what the NIMH described as an 'exhaustive review' of Breuning's work at Pittsburgh." After this investigation, and Breuning's guilty plea in federal court, the University of Pittsburgh paid back over $163,000 in grant money the agency had given the university for Breuning's research. Although the NIMH panelists praised Sprague for telling them about Breuning, they criticized him for not properly overseeing Breuning in the first place.

Retribution for Whistle-Blowers?

Part of Dr. Sprague's testimony to Congress reflected poorly on the University of Pittsburgh. A report on science fraud by Congressman Weiss's subcommittee stated, the "University of Pittsburgh Medical School's record of dealing with allegations of scientific misconduct is consistently inadequate."[16] As if to underscore this, Sprague received a July 12, 1989, letter from George A. Huber, Vice President and Counsel, University of Pittsburgh, Medical and Health Care Division, saying that Sprague's written and oral testimony concerning the University of Pittsburgh was "inaccurate and untrue" and also "slanderous and libelous." Huber insisted, in underlined and capital letters, that his university did "*NOT*" attempt "to cover up" Breuning's bad behavior.

Not only did Huber "demand" that Sprague "cease" saying what he had been testifying to about the University of Pittsburgh, and "desist" from saying this sort of thing in the future, but also

Huber wrote, "I insist" that Sprague "retract" all the "inaccurate" parts of his testimony to Congressman Robert Roe's subcommittee, where Sprague had also testified. Huber also insisted that Sprague explain his "misstatements" to the Roe subcommittee.

Huber then stated that unless Sprague ended at once making the sorts of statements he had made in his sworn testimony, the University would "take whatever lawful corrective measures are necessary" to stop him. Specifically, Huber wrote that the University of Pittsburgh would file "legal action."[17]

Mr. Huber may not have noticed the comments of John Dingell when Sprague testified to Dingell's subcommittee: "We'd like you to know, Doctor, that we have a continuing interest in your well being, and that as you, Dr. Sprague, go about your business, if you have any reason to feel that there's any retribution taken against you as a result of your appearance here today, or your assistance to the committee, we'd like to know about it punctually so that we can call before the committee those individuals who might have such interests to express their reasons."[18]

Dingell wrote an appropriate letter to Wesley W. Posvar, president of the University of Pittsburgh. Posvar wrote back to Sprague that even though the Medical School counsel said he was writing "on behalf of" the university, in fact, Posvar and Lewis Popper, the university counsel had not been aware of Huber's letter until Dingell wrote them. Posvar wrote that the Medical School counsel's accusations were "intemperate," even though Posvar himself could not find any evidence of a cover-up at the University of Pittsburgh. But Posvar acknowledged that the university had been "slow in investigating" the total amount of the fake research.[19]

Dingell's top investigator, Peter Stockton said, "There's no incentive for the University to do the right thing. The chances of being caught covering up are maybe 1%. But if you do the right thing, there's almost a certainty you'll be embarrassed. So if you cover-up, you get away with it. But on the off chance that you do get caught, and the NIH [National Institutes of Health] releases the information, you get embarrassed, but you don't' lose any money."[20]

The University of Pittsburgh was not the only institution to take action against Dr. Sprague after he blew the whistle on Dr. Breuning. After nearly 18 straight years of funding, the National Institute of Mental Health [NIMH] canceled Dr. Sprague's grant. "I got a letter early in March that it was deferred, so there was no funding, so my research was temporarily halted. Subsequently later in the summer, the grant was renewed," testified Sprague.[21] But the renewal was for only one year, at 29% of the amount recommended by the study section of his peers "to analyze the data and wind up things," said Sprague.[22]

According the Weiss Subcommittee Report on Scientific Misconduct, "Many of the NIMH review documents have been destroyed, reportedly as part of the agency's usual procedures, so that there is insufficient documentation to prove whether or not the NIMH's actions regarding Dr. Sprague's grant is in any way related to the fraud case involving Dr. Stephen Breuning."[23]

Dingell asked Sprague, "Was that a regular NIMH practice to cut all grants to 1 year, or did they just cut your grant to 1 year?"

Sprague replied, "No. That can happen. I think it's not too frequently that that happens, but it can happen."[24] A few months later, when Sprague was a panelist at a symposium on research fraud, a person who was not a panelist, Michele Applegate, the misconduct policy officer for the federal agency that includes the NIMH, made arrangements with the Chair of the session to speak. She stood up in the audience, introduced herself to those attending and said, "I feel that I need to respond to some comments that Dr. Sprague has raised. . . . I will categorically state . . . that there was no retribution." She also stated, contrary to the facts, ". . . it wasn't that people were being given drugs that were harmful to them. . . ."[25]

The message of the case is clear; there is an extremely good chance that pulling off a scientific fraud pays, and whistle-blowing almost certainly does not pay off. Why didn't other scientists step forward? Sprague recalled that of Breuning's thirty coauthors, "only one, C. Thomas Gualtieri, took a public position in support of me."

At a conference at the Massachusetts Institute of Technology (MIT), titled "Error, Fraud and Misconduct in Science," MIT Professor of Biology Frank Solomon explained, "I had a colleague say to me the other day, 'No group of colleagues will ever find a colleague guilty of malfeasance. It's too difficult for them to do it.' And when I asked why, he smiled and said, 'We all have skeletons.' I knew I had a skeleton; he meant in the closet. I went and checked, and I'm clean, which is why I can come here today."[26]

THE DARSEE CASE

Another example of the response within the scientific community to evidence of fraud is the Darsee case, which I will mention only briefly.

Dr. Eugene Braunwald, of Harvard University, had been Dr. John Darsee's mentor at the Harvard Medical School when three co-workers saw Darsee fake the data for an experiment. They told Braunwald they thought Darsee had faked the data on many of his other experiments, which had resulted in close to 100 published papers over a 2-year period, quite a few of them coauthored with Braunwald. Braunwald didn't buy it. He and another scientist secretly looked into Darsee's work for the next five months and found nothing wrong. They made no public disclosure of their investigation. Braunwald said, "He clearly was one of the most outstanding, or the most outstanding, of the hundred and thirty research fellows I have been privileged to work with. Public disclosure would have ruined him for life."[27] Darsee admitted the one act of fraud that had actually been observed, and although he had to give up his Harvard job, he was allowed to keep working in the laboratory on further experiments. In fact he kept at it during the five-month investigation.

An NIH investigation found material that failed to add up in some of Darsee's results in an NIH-funded study spanning four different universities. At that point Harvard acknowledged to the NIH that Darsee had already admitted to a single instance of fraud. Harvard then set up a committee of five professors to investigate him. They concluded that there was nothing fraudulent in any of the published research reports based on work done at Harvard.

The NIH set up its own committee, which came to a radically different conclusion: Nearly every paper Darsee had produced was fabricated. As the Weiss subcommittee report stated, "The committee criticized Harvard's handling of the investigation, as well as the lax supervision and poor record keeping in the lab, which was under the direction of one of the nation's leading cardiologists."[28]

As this case illustrates, top scientists tend to glide through these embarrassing episodes.

THE BALTIMORE CASE

Without question, the most notorious case of misconduct in basic research in recent memory is what has become known as the "David Baltimore case." It involves the charge, now preliminarily accepted by the NIH in a draft report, that key data were faked in a supposedly pathfinding paper on immunological research, which had been coauthored by Dr. Baltimore and five others and published in 1986, in the journal *Cell.*[29] The paper has since been retracted, at Dr. Baltimore's request.

On December 2, 1991, Dr. Baltimore announced his resignation as president of Rockefeller University, having only taken up the post 18 months earlier, on July 1, 1990. He explained in his letter of resignation that the episode surrounding the paper had produced a "climate of unhappiness" at Rockefeller "that could not be dispelled."[30] Before taking up the presidency at Rockefeller University, and at the time of most of the events described in this chapter, 1985–1990, he was the head of the Whitehead Institute, a private biomedical research institute affiliated with MIT. Winner of the Nobel Prize in 1975, along with Howard M. Temin, for a discovery that led to the understanding of how cancer can be induced by viruses, Dr. Baltimore had been viewed, before these events, as a role model for thousands of scientists throughout the world. After these events, his public image suffered considerably. *TIME* magazine, in its issue of January 6, 1992, gave its opinion of the best and worst in science in 1991. The magazine listed nine examples of the best, but only one of the worst—David Baltimore for "his prideful refusal to re-examine the data." The magazine also said he "belittled the whistleblowers and decried the government's 'witch hunt.'"

The case is extraordinary because not only did it involve a particularly celebrated Nobel Prize winner, it also involved two NIH investigations, three hearings by Congressman Dingell's Oversight and Investigations Subcommittee, and two investigations by the forensics laboratory of the U.S. Secret Service. Perhaps most extraordinary is that it is known as the "Baltimore case" at all. Baltimore was never charged with faking data. Were it not for his own deliberate efforts to make this *his* case, the investigation unquestionably would have been hardly noticed. It would have been known as the "Imanishi-Kari case," after Dr. Tereza Imanishi-Kari, head of an MIT laboratory where the original research on transplanted genes was alleged to have taken place. She was one of the principal coauthors of the paper and was the person actually charged by the NIH with faking the data.

Briefly, the basic elements of the case are as follows: Postdoctoral researcher Margot O'Toole, working in Dr. Imanishi-Kari's laboratory, was unable to replicate the key result described in the *Cell* paper. Dr. O'Toole consulted a laboratory notebook that was up on a shelf in plain sight; it had been pointed out to her and left at her disposal by another postdoctoral researcher who had previously been in charge of the notebook and had returned to her native Brazil. O'Toole noticed that 17 pages in this notebook apparently described the original key experiment but clearly showed that the result reported in *Cell* had not been obtained. O'Toole brought the matter up with scientists with whom she was, at the time, close. This resulted in peer review "investigations" first at Tufts University and then at MIT, the latter with the active participation of Dr. Baltimore. In their reports, both university inquiries dismissed O'Toole's concerns, and her career was essentially derailed. Science fraud hunters Dr. Ned Feder and Walter Stewart contacted Dr. O'Toole and brought the case back to life, resulting in the series of investigations by the NIH, Congressman Dingell, and the Secret Service, and ultimately the explosion of the case onto the front page and the evening news.

A number of the most important people in American science condemned the search for truth in the Baltimore case. They wanted to ignore the question of fraud and let the so-called replication process sort out whether the central claim of the *Cell* paper was true or not. Believing that science is inherently self-correcting, they were convinced that the truth would ultimately win out. It is an approach that pretends to seek the truth, but in fact rewards lies.

The case documents several current attitudes widely shared within the scientific community. One is that scientists treat raw scientific data, even when generated by public money, as private property. As a result they are reluctant to examine data when somebody charges error or misconduct.

A key element of the scientific method is the attempt to publish results that contradict previously published results. The whistle-blower in the case, Dr. Margot O'Toole has testified under oath that she did not do this because at a key meeting, "Dr. Baltimore told me that he would say I was wrong, it didn't seem that it would do any good because I thought he would be believed and I would not be, and I think that point has been proven pretty conclusively."[31] When NIH scientists and science fraud sleuths Walter Stewart and Ned Feder tried to publish a dissenting analysis of the conclusions in the disputed *Cell* paper, based on questions about the raw data, the scientific establishment again closed ranks. As O'Toole testified, "The authors [of the *Cell* paper] didn't want to go over the data and the editors didn't want to publish the manuscript."[32] For a while, Stewart and Feder's bosses at the NIH ordered them not to further submit their paper for publication, and ultimately, no journal would publish it.

One reason lies can be rewarded in science is that some major researchers in the United States, judging from their public comments, do not always fully advocate the scientific method. David Baltimore himself has taken both sides of the question of whether he is a stickler for rigorous record keeping. He has testified: "I realize that with the increased sensitivity about scientific fraud, it is more crucial than ever that scientists keep records that can easily be referred to so that the scientific basis for published results is readily determined at any time."[33]

At the same congressional hearing, however, he also testified: "[T]he Secret Service analysis, as elegant as it is, it seems to me simply proves that different scientists do their science in different ways. I guess I encourage that. I encourage scientists to be individuals, and I'm less worried about somebody who is messy, than I am about somebody who is not smart. . . . I think we should find ways to encourage individuality and not to straightjacket science with a preconceived notion of how notebooks should be kept, or how anything about science should be done."

Presumably based on this view, Baltimore testified about the record keeping of the person subsequently charged in an Office of Scientific Integrity (OSI) draft report with faking data, Dr. Imanishi-Kari: "I know full well from my interactions with her, that with her understanding of the experimental system, which is quite remarkable, that she could reconstruct the history of her work in great detail and with the kinds of cross-checks in it that indicate that she's not mis-remembering things, that she has it all under control."[34]

DAVID BALTIMORE'S LETTER TO THE SCIENTIFIC COMMUNITY

The angry reaction from the scientific community to Dingell's investigation of the Baltimore case was solicited in two "Dear Colleague" letters, one from Dr. Baltimore, and a second from Philip A. Sharp, the director of MIT's Center for Cancer Research. The more important letter, which to an enormous degree explains why the case became known as the Baltimore Case instead of the Imanishi-Kari Case, was written by David Baltimore on May 19, 1988. The nine-page document was sent nationwide on Whitehead Institute stationery, where, as the letterhead explained, Baltimore was the director. According to Walter Stewart, interviewed in May 1989, "Baltimore is considered a national treasure, the moral leader and role model for thousands of scientists. They view him as top of the line. When Ned Feder and I pointed to other, earlier cases of fraud in science, they would say, 'Maybe, but that's not the *best* science, that's not David Baltimore.'"[35]

The occasion for Baltimore's letter was Dingell's April 12, 1988, hearing at which a number of people testified, including Margot O'Toole, Ned Feder, Walter Stewart, and an extraordinary eyewitness to the events, Charles Maplethorpe, whose testimony was compelled against his will by subpoena. Neither Baltimore nor Imanishi-Kari testified, and Baltimore protested the fact that he hadn't been invited in his letter, in which he also said that he and the other coauthors were not "even informed" about Dingell's investigation.

Baltimore wrote that one purpose of his letter was to "set the record straight." He noted that there had been broad but often shallow press coverage, which in some cases had been "downright wrong." His letter would serve to clear his own name and those of his coauthors, "who have been compromised by this attack." Then he added what he described as a "more compelling reason" for his letter: "A small group of outsiders, in the name of redressing an imagined wrong, would use this once-small, normal scientific dispute to catalyze the introduction of new laws and regulations that I believe could cripple American science."[36]

Stewart disputed Baltimore's assertion. "Dingell picked this case up simply because it is an interesting test case of what is happening," he said. "It's a typical case because the facts are complex, it involves powerful people to whom junior people pointed out mistakes and that was followed by an elaborate and concerted coverup."

Dingell himself addressed the question of his investigation in the opening remarks of his second set of hearings on the case in May 1989: "The Congress authorizes approximately $8 billion for NIH alone. It is the responsibility of the Committee to assure that this money is properly spent and that research institutions, including the NIH, which receive these funds, behave properly. . . . critics have claimed that the Congress is not capable of understanding science, or even raising questions about science.

"However, no one questions the ability of the Congress to deal with these issues when Utah scientists demand $25 million for a cold fusion experiment, or the Air Force needs $70 billion for the Stealth Bomber Program, or hundreds of billions of dollars are requested for Star Wars, or when enormous sums of money are requested for enormous particle accelerator programs. . . ."[37]

At the end of the hearing, Dingell added: "The Chair and this committee and this subcommittee have spent an enormous amount of time protecting science, protecting the rights of individuals, protecting the programs of the NIH and Federal funding for Federal research. . . . We have no desire to intrude into the laboratories and we have no desire to dictate the course of scientific research. That is not our goal.

"Our purpose is quite different. Our purpose is to see to it that where questions of the sort that we are confronting today arise that they are properly addressed and that a proper mechanism exists to address those and that that proper mechanism functions fully, correctly and properly."

Baltimore's letter generated such an extraordinary and ill-informed response from the scientific community that Margot O'Toole would write, after the NIH had branded key parts of the disputed paper as faked, "He now owes an apology to Congressman Dingell, to scientists who attacked Dingell based on Dr. Baltimore's personal and professional assurance that the paper was fine, and to the scientific community as a whole because of the damage this has done to the reputation of the profession."[38]

The most serious discrepancy concerns Margot O'Toole's objections to the basic conclusion of the *Cell* paper. Baltimore wrote: "Early in 1986, after the *Cell* manuscript had been submitted, Dr. Margot O'Toole, a postdoctoral fellow in Dr. Imanishi-Kari's laboratory, *questioned the interpretation* [emphasis added] of some of the data emanating from that laboratory. . . . Dr. O'Toole's criticisms, as described by her, were based on certain experimental attempts of hers involving some of the reagents used in the study described in *Cell.* A part of her criticism was based on 17 pages of selected laboratory notes, a small fraction of the notes compiled during this project. She summarizes her criticisms in the first paragraph of her letter [to an MIT scientist looking into the matter]. . . ."

Any scientist reading the preceding paragraph from Baltimore's letter, and knowing nothing else about the matter, other than that some sort of controversy exists, would undoubtedly, and erroneously, conclude that O'Toole was offering "an alternative interpretation of the data."

After the NIH's 1991 draft report became public, and Baltimore called for the *Cell* paper's retraction, O'Toole herself characterized her memo to the MIT investigator, Dr. Herman Eisen: "My June 6, 1986 memo stated that experiments described in the paper had not actually been performed, and that other experiments described had not yielded the claimed results."[39] There is some question as to whether Baltimore had already expressed O'Toole's true concerns in a letter to Ned Feder and Walter Stewart more than a year before. In the letter, dated January 21, 1987, in which Baltimore referred to O'Toole as a "discontented postdoctoral fellow," he stated that he had been "aware for some time" that O'Toole had "raised questions about some of the data" in the *Cell* paper. "I have satisfied myself that within the norms of scientific evidence the data are correct," Baltimore wrote. However, his letter to Feder and Stewart also gave equal weight to the interpretation of O'Toole's position he would later take in his letter to the scientific community. He wrote that the data was "consistent with" the interpretation of it in the paper but he was aware that other researchers "can interpret the data differently."[40]

O'Toole's written testimony at the April 12, 1988, Dingell hearing describes what happened after she presented her concerns to a Tufts scientist, Dr. Henry Wortis, who was Imanishi-Kari's friend and colleague. As agreed in advance with O'Toole, Wortis brought up the disputed data with Imanishi-Kari. "Within half an hour after the meeting had been scheduled to begin, Dr. Imanishi-Kari called me. She accused me of vindictive motive and threatened to sue me."[41]

The entire thrust of O'Toole's testimony at the 1988 Dingell hearing, given more than a month before Baltimore's "Dear Colleague" letter, is that she was concerned about possible faked data, or a lack of appropriate data at all. O'Toole submitted for the congressional record a written chronology of events after she started working in Imanishi-Kari's MIT laboratory as a postdoctoral fellow:

> *July and August 1985.* Some of my experiments yielded results that contradicted Dr. Imanishi-Kari's claims. . . . I reported to Dr. Wortis, my former advisor and a close associate of Dr. Imanishi-Kari's, that I considered many of the procedures in the laboratory unsound and improper. . . .
>
> *November 1985–March 1986.* . . . Despite many attempts, I was unable to replicate the results of an important experiment Dr. Reis [an earlier researcher in the lab, now in Brazil] and I had done in September. When subsequent experiments continued to contradict my original results, Dr. Imanishi-Kari decided that she herself would perform these experiments. She told me to confine my laboratory duties to mouse husbandry. In March, I discovered that Dr. Reis had made an error which invalidated our initial results and explained the failure to replicate the finding. Dr. Imanishi-Kari, however, still insisted that she would eventually repeat my initial results and she continued to do the experiments. . . .
>
> Dr. Imanishi-Kari told me to prepare for publication a study I had done that was closely related to that described in the Weaver et al. paper. She gave me instructions to omit certain facts about my results. This would have resulted in a misleading paper. I pointed this out a number of times to her but she persisted in her instructions. I eventually resolved to simply stop arguing and just not do what she told me. . . . I gave my notice but agreed to continue taking care of the mice through May.
>
> *April 1986.* Dr. Imanishi-Kari showed me results of her first cell transfer experiment. By manipulating the data, she forced the results to conform to her hypothesis. She pointed out that, if my data were manipulated in the same way, they also would support her hypothesis. She said that I was too "nit-picky" and that "better reagents would confirm her results."

What was O'Toole's "nit-picky" attitude about? O'Toole had some frozen blood from the earlier December 1985 experiment she had done with Dr. Reis, but she couldn't replicate the earlier result. So Imanishi-Kari tried the experiment herself. Walter Stewart explained what happened: "The results of the experiment in question ended up as a series of five numerical values for the experimental group and five for the control group. You then average each. If they are different, you have an effect. Margot's hadn't differed. Thereza's hadn't differed either. Thereza then threw out the extreme values at one end for the experimental group and the extreme values at the other end for the control group. Then she could show an effect.* Margot had thought this was harebrained, and so had resolved by April of '86 to get out of that lab, because the science there was bad."

According to Stewart, at this point, "Margot still didn't suspect fraud, just incompetence." In a January 1992 telephone interview, Margot O'Toole, although fully corroborating Stewart's account, added that she had no independent evidence to support the story and that Dr. Imanishi-Kari disputes it. Indeed, Dr. Imanishi-Kari denies all charges of wrongdoing and misconduct in this case.

O'Toole's testimonial chronology continued:

> Dr. Imanishi-Kari admitted to me that one of the graphs in the Weaver et al. paper was misrepresented and that the true results agreed with mine. This startling disclosure, coupled with my increasing knowledge of the available laboratory reagents and procedures, led me to doubt that the experiments could have been performed as described in the Weaver et al. manuscript. On two occasions I informed Dr. Weaver that I was concerned about the results reported in his paper. On both occasions, however, his schedule did not permit a thorough scientific discussion. The Weaver et al. study appeared in print. . . .
>
> *May 7th.* . . . While examining the [mouse] breeding records Dr. Reis left in my care, I recognized that some of the entries were not breeding records, but the results of some critical

* Try this yourself. Write two columns of the numbers 1, 2, 3, 4, 5. Label one "control" the other "experimental." Now cross out the 1 and 2 under "control," and then the 4 and 5 under "experimental." You have left "3, 4, 5" under "control" but "1, 2, 3" under "experimental." If these were reports of lab results, you would now clearly have a difference.

experiments for the Weaver et al. paper. In order to study these records more carefully, I xeroxed the notes and examined them during my lunch break. I became convinced that several of the major assertions of the paper were actually contradicted by the experimental results. Moreover, certain crucial facts, much like the facts she had instructed me to omit from my manuscript, were also omitted in the Weaver et al. study. I then approached Dr. Imanishi-Kari and, without mentioning the notes I had xeroxed, asked if I could examine the records for a related set of experiments that she said she had performed herself. . . . She searched the laboratory for some time. Not only did she not find the records I requested, she was unable to find any of her records concerning the claims in the Weaver et al. paper.

Although Imanishi-Kari had beaten out nearly 100 other scientists for the job at MIT five years earlier, she had not been given tenure and had applied for a job at Tufts. According to Stewart, "Her major paper, the disputed one published in *Cell,* had come just at the right time." This brought researchers at Tufts into the case, because they were considering her for a job, and the decision was not yet final.

O'TOOLE'S 1988 TESTIMONY AND THE TUFTS "INVESTIGATIONS"

O'Toole's chronology recounts a conversation she had with one of the Tufts scientists, Dr. Brigitte Huber after the first meeting of the Tufts inquiry:

May 17th. Dr. Huber reported to me that . . . Dr. Imanishi-Kari was unable to produce most of the relevant records. She described some records that they did examine and agreed with me that these tended to support my objections and actually confirmed one of my most important assertions.

Dr. Huber disputed this statement. In her sworn testimony to the Dingell subcommittee, Huber wrote, "We concluded that there was no discrepancy from the published data, and that the allegations had been based on a misunderstanding."[42] Tufts scientist Dr. Henry Wortis, who conducted the Tufts inquiry involving two meetings, subsequently wrote a June 4, 1986, report on it that concluded, in capital letters, "NO EVIDENCE OF DELIBERATE FALSIFICATION. NO EVIDENCE OF DELIBERATE MISREPRESENTATION. ALTERNATIVE INTERPRETATIONS OF THE EXISTING DATA CAN BE MADE, BUT THAT IS THE STUFF OF SCIENCE."[43]

Another scientist present at the first meeting, Robert T. Woodland, PhD, wrote Wortis on June 10, 1987, that his report "accurately and fully" described the meeting and its conclusions. Woodland, who was not present at the second meeting of the Wortis panel, added that Wortis's report omitted a point that perhaps it would have been "reasonable" to include. This omission was the "fact" that the members of the inquiry panel not only "examined" Imanishi-Kari's original laboratory notebooks but also "verified all conclusions from the primary data." The letter also mentioned Imanishi-Kari's cooperative, "full and open manner" while the panel, according to the letter, went through the data.

However, after Wortis agreed with Dingell that the data for at least one key data point were not in the lab notes and therefore could not be used to verify the conclusion, he testified, "I don't think we were asked, nor did we review in 1986, you know, every last data point. I don't think that's what we were asked to do." As for what Imanishi-Kari did at the panel meetings, Wortis testified about a panel meeting, which had just been specifically identified by Huber as the first meeting, "[W]e said we would like to see that data and at that point she began to cry and she said, you don't believe me? You don't trust me? We said no, we want to see the data and she got out the data and we looked at it then."[44]

Wortis also testified about the first meeting of his panel, "At the conclusion of the meeting, we all agreed that there were no significant problems with the paper. I believe it was Dr. Huber who contacted Dr. O'Toole to report back on our meeting. To our surprise, Dr. O'Toole was not satisfied with our conclusions and reiterated her concern that the underlying data did not support the conclusions of the *Cell* paper."[45]

Neither Huber, Woodland, nor Wortis agreed with O'Toole's testimony about their meeting. Wortis, however, agreed that O'Toole was concerned *not* about varying scientific interpretations, but about *whether or not the underlying data supported the paper.* Woodland emphasized in his letter that they had looked at "the primary data"—which is surprising since the NIH was later unable to do the same thing. And Huber testified that they had looked to see if there was a "discrepancy from the published data."

O'Toole's dissatisfaction with the outcome of the meeting led to another meeting. O'Toole described the second meeting in her oral testimony: "Dr. Imanishi-Kari and I reviewed the data. My assertions that the data did not support the published claims were completely confirmed. I left that meeting under the impression that the paper would be retracted.

"The next day, however, I was informed by Dr. Huber that, although she and Dr. Wortis agreed with me scientifically, they felt that fraud was not the cause of the discrepancies. They had decided therefore that since a retraction or correction might be very detrimental to Dr. Imanishi-Kari, it was best for all concerned to drop the matter. I disagreed, saying that other laboratories were relying on the data and that we had a professional responsibility to make the truth known."

Wortis's testimony completely contradicted O'Toole's. He testified that after this second meeting: "I believed we were all satisfied at that time that the underlying data supported the conclusions of the paper. There was no evidence of falsification or misrepresentation. There was no discussion of a retraction or clarification. Of course, alternative interpretations of the existing data can be made, but as we stated, 'that is the stuff of science.'"[45]

Huber's testimony also contradicted that of O'Toole: "At the meeting we carefully went through the primary data provided by Dr. Imanishi-Kari, and we came to the conclusion that there may be a minor problem with one of the Figures in the published report. We were convinced that even the minor inconsistencies detected had no impact on the validity of the reported results. The data certainly did not need to be retracted. Based on Dr. O'Toole's expressed concerns at that time, we found no other problems. I felt and still feel that we thoroughly addressed Dr. O'Toole's concerns."[46]

Although both Wortis and Huber contradicted O'Toole, their testimony seems clearly to show that O'Toole was concerned with the data, not with an alternative interpretation.

Baltimore's "Dear Colleague" letter, delivered after O'Toole's testimony, clearly did not state what she was worried about, or why Congressman Dingell was holding hearings into the case.

BALTIMORE'S PARTICIPATION IN THE INQUIRY

After the second meeting of the Wortis committee, Dr. Herman Eisen of MIT held a meeting in June 1986 to look into the matter. It was attended by Baltimore, Weaver, Imanishi-Kari, and O'Toole. Baltimore testified about it to the Dingell subcommittee at his May 4, 1989, appearance, during Dingell's second set of hearings on the case: "I became aware of Dr. O'Toole's criticisms only a few days previous to the meeting when Dr. Eisen called to explain the situation. . . . The June 1986 meeting with Dr. Eisen was my first and last encounter with Dr. O'Toole over this issue. Her four points were carefully considered, with my role being either bystander or occasional questioner. I left that meeting convinced that there was no serious error in the *Cell* paper, but was also impressed by the quality of Dr. O'Toole's analysis, and believed that further experimentation might validate certain of her points, and that therefore certain of her criticisms could be a valuable contribution to science."[47]

In his "Dear Colleague" letter, Baltimore briefly referred to the Eisen meeting, and quoted a letter from Dr. Eisen extensively. Eisen's letter stated that no one at the meeting thought O'Toole's disagreements were "frivolous." Her objections were grounded on "pretty carefully thought out ideas of the literature of the analytical method." Not only are such disagreements "not uncommon" throughout science in general, they are "plentiful in Immunology." The traditional way to resolve them, said Eisen, is through publication, followed by evaluation and possible attempts at replication by scientists at other laboratories. "This is the way science operates," wrote Eisen, adding that this sort of "contentiousness" even "helps drive the science 'engine.'" The only "unusual" part to what Eisen described as the "exercise" is the "intensity of the feelings generated and the circumstances concerning Dr. Imanishi-Kari's pending appointment at Tufts." Although Eisen concluded that O'Toole had found "an error," it was "not a flagrant error." It was "too minor to warrant a retraction." Eisen concluded, that the other issues were primarily "matters of interpretation and judgment," that he felt should be handled by letting the scientific process "take its course."[48]

O'Toole felt that her concerns raised at the MIT meeting were "well founded. None of the objections were answered and I made this point clear a number of times. Dr. Imanishi-Kari told Dr. Baltimore that she had not brought the critical records to the meeting because she had not thought she would need them and because they were written in Portuguese. Dr. Baltimore assured her that he did not need to see any other data."

In her direct testimony, O'Toole elaborated on the data she showed Dr. Baltimore: "He said you couldn't tell anything from it [the data]. But he had published—he had his name on a paper that made these extraordinary claims, based on that paper, and he was admitting that you couldn't tell anything at all from the data.

"And I agreed entirely. You can't tell. You can't make those claims from that data. . . . Dr. Imanishi-Kari confirmed without dispute that she had misrepresented the data in figure 1. Dr. Baltimore asked her why she had done this and she said she thought they must have 'got that result once.' He said that was not acceptable and he would talk to her in private. He outlined a series of new experiments she could do that could, if successful, support her hypothesis. He said he thought that should satisfy me. He strongly advised me to drop the matter 'for my own good' and told me that a lot of his time went into calming people like me down. . . . He said that my only recourse, as I saw it, was to write to the journal. He added that if I did this he would personally write a rebuttal 'similar to the recent one in *Nature.*' I stated that I considered my responsibilities discharged and intended to drop the matter. Dr. Baltimore said, 'Oh, well, then there is no problem.' He stated that he had reason to believe that part of the study was true and that therefore no correction was warranted. He asked Dr. Eisen to write up a memorandum stating that the matter had been 'aired.'"[49]

In his oral testimony to the Dingell subcommittee, Baltimore said, "Let me say that I believe that what Dr. O'Toole did up to this point was healthy and proper and should be enthusiastically encouraged. But at this juncture, what was needed to decide the merits of her arguments were new experiments, and those were not done."

In his written testimony, Dr. Baltimore elaborated: "At the end of the June 1986 meeting I believed that her criticisms had received a fair and useful airing. . . . I felt, however, that Dr. O'Toole was not satisfied with the outcome; she made it clear that she demanded a published recognition of the veracity of her interpretation. I did not realize that having failed to get the desired outcome from two peer reviews, Dr. O'Toole would continue to press her case in additional and less conventional forums."

One of the most striking testimonies during the Dingell subcommittee hearings came from Charles Maplethorpe, MD, who was forced to testify under subpoena. Although Maplethorpe already had a doctoral degree in medicine, he had been in MIT's PhD program and was assigned to Imanishi-Kari's laboratory. He was suspicious of Dr. Imanishi-Kari because she wouldn't share her data with others, including himself. He testified that he overheard a conversation between Dr. Weaver, a coauthor of the paper and a Baltimore protégé, and Dr. Imanishi-Kari in June of 1985, before the publication of the *Cell* paper, and a few days after the first public presentation of the data in a seminar given by Dr. Imanishi-Kari.

"[The conversation] took place in the evening in the laboratory. My impression is that Dr. Weaver was coming to the laboratory with a friend to speak with Dr. Imanishi-Kari about her data. And my feeling was that they had personal doubts about it, but they wanted to discuss it with her in a friendly way, in a way that would not arouse her suspicions.

"I was working in the lab at that time trying to finish for my thesis defense in August. But nevertheless, I was paying attention to the conversation and, at that time, I heard Dr. Imanishi-Kari tell Dr. Weaver that she had some problems with this reagent called Bet-1, which is the reagent that Dr. Margot O'Toole had problems with later.

"And what I heard Dr. Imanishi-Kari tell Dr. Weaver was that she was obtaining the same results that Dr. O'Toole subsequently obtained."

Congressman Ron Wyden asked Maplethorpe at the subcommittee investigation, "And how were these issues resolved prior to publication by Dr. Baltimore, Dr. Imanishi-Kari, and others?"

Maplethorpe replied, "Well, the issues that I heard them discuss were not resolved."

Wyden asked, "What was your reaction when the *Cell* article appeared with these various concerns unresolved? You were shocked? Were you not?"

Maplethorpe answered, "Well, when a train wreck happens, one is shocked, but if one sees the train wreck about to happen, I mean. . . ."[50]

Maplethorpe testified that he spoke to the assistant to the president of MIT about his concerns but in response was merely given a photocopy of the MIT fraud guidelines.

Wyden asked Maplethorpe, "And that was the extent of MIT's response?"

"As far as I know," testified Maplethorpe.

"Did you feel that had you pressed your concerns with MIT that perhaps that would have worked to your disadvantage with respect to your PhD and academic situation?"

Maplethorpe answered, "I felt there was no question, but that if I were to make a formal charge of fraud, that it would not be taken seriously and then I would be the person who would be worse off for it."

BALTIMORE'S MEMO TO EISEN

In David Baltimore's letter to the scientific community, he alludes to a situation he wrote about in a memo to Dr. Eisen, "[F]or a short period of time in September, 1986, a misunderstanding led Dr. Eisen and I to doubt the specificity of the Bet-1 reagent whereas it was only the case that occasional preparations were not specific."

Whether Baltimore's allusion to the memo was an attempt to inoculate the scientific community to its contents, which were not revealed until a year later, is uncertain; but the memo was, in fact, nothing short of sensational. It stated that after thinking heavily about the "situation" that arose from his coauthoring the disputed paper with Imanishi-Kari, his views had "jelled." He then spelled out his opinions in a series of numbered points. The first of these stated that "The evidence . . . is clear" that a key antibody, Bet-1, "doesn't do as described in the paper." This was a major element in O'Toole's objections. He referred to a statement, presumably from an earlier communication with Eisen, that Imanishi-Kari had told Eisen that she "knew it all the time" concerning the problems with the Bet-1 antibody. Baltimore characterized this reported statement by Imanishi-Kari as "a remarkable admission of guilt." Baltimore wrote that he had had no idea, nor did coauthor David Weaver, that this antibody resulted in an "ambiguity." His first point closed with Baltimore saying that he genuinely could not understand the reason that Imanishi-Kari made the decision to employ this ambiguous data, which "mislead both of us and those who read the paper."

The next paragraph states that the analysis based on the unreliable antibody is "meaningless." However, it is only of consequence for quantitative exactness. Qualitatively, the analysis is "OK."

His third point argues that a retraction would particularly harm David Weaver because he is listed on the paper as the first author, the portion of the paper he contributed "is what makes the paper strong," and he had "nothing to do with" the bad data. He had accepted that data "in good faith." Based on this, Baltimore "would hate" to watch as Weaver's integrity was thrown into doubt. Baltimore wrote that he held this view even though in a coauthored manuscript all the authors are obliged to assume "responsibility." As a result, all the authors, clearly including himself, are "in a sense culpable."

His fourth point continued the discussion of what a retraction would mean, and how relatively rare they are even though many articles and portions of articles in the scientific literature later prove to be "wrong" but are not publicly pointed out as such. A retraction "goes to the heart of a paper and implies that the data is generally unreliable." Baltimore was confident that Weaver's work in the paper is "solid." Had all the work come out of the laboratory of Imanishi-Kari, Baltimore would ask himself "what else might be wrong."

He concluded that a retraction based on the Bet-1 matter would cause scientists to question other, presumably Weaver's, "quite solid work." It would also "harm the innocent," presumably Weaver again, although this could apply equally to the other coauthors except for Imanishi-Kari. Baltimore added that some sort of acknowledgment—he didn't specify the details—should be made to colleagues concerning the unreliability of Bet-1. He also added that from then on, he would be "skeptical" of work by Imanishi-Kari.[51]

Baltimore explained himself at the 1989 Dingell hearing in this way: "Professor Imanishi-Kari is also a victim. English is her fourth language. She has difficulty communicating in English. . . . It was, in fact, the difficulty communicating with Dr. Imanishi-Kari that gave rise to a misunderstanding between Dr. Eisen and myself, that, in turn, led to a letter from me to him.

A letter of which I am not proud, and a letter which can easily be misconstrued.

"In September, 1986, Dr. Eisen had a chance conversation with Dr. Imanishi-Kari, and thought he heard her say that a reagent used in the study didn't work. He told me about it. Instead of doing what I should have done, and that is to call Dr. Imanishi-Kari and talk it over with her further, I fired off a letter to Dr. Eisen based only on this misunderstood statement, and suggesting that since the supposedly faulty reagent wouldn't change the paper's conclusion, we not formally retract the paper, but only acknowledge the error to fellow scientists."

Although Baltimore testified that he "fired off a letter," the actual letter states: "After much thought about the situation brought on by my collaboration with Thereza Imanishi-Kari. . . ." This seems to suggest that there was more to the issue than a single Eisen conversation. Nonetheless, the subcommittee did not follow up on the matter. As *Science & Government Report* put it, Baltimore "smoothly sailed past a great potential embarrassment."

Baltimore went on to testify, "Dr. Eisen spoke to me within a day or two and explained that he had misunderstood Dr. Imanishi-Kari. The reagent was very well suited to its task." Baltimore added under questioning from Dingell, "That letter was completely inoperative in its significance within a couple of days of its writing."

Eisen corroborated Baltimore's testimony. He testified that he accidentally ran into Imanishi-Kari: "It was one of those corridor conversations where people say things in passing but I was bothered by it. I communicated it to David Baltimore. He was upset and you see the result of that in that letter to me. When I got back to Dr. Imanishi-Kari, she explained much more about it. . . ."[52]

STEWART & FEDER

The reason Baltimore was testifying to the Dingell subcommittee at all was due entirely to events that had happened almost three years earlier, after the June 1986 MIT–Eisen committee meeting at which Baltimore had played a key role. Following that meeting, Margot O'Toole was thoroughly ground down and wanted to drop the whole matter. She had struggled to complete a PhD but was now branded by the leading citizens in the small world of immunological research. Not only was she on her way out of Imanishi-Kari's lab, but she couldn't get a recommendation to work in any other lab. Her brother hired her to do office work in his moving company.

At this point, Maplethorpe read a newspaper article about Walter Stewart and Ned Feder, and two tenured NIH scientists who had recently devoted themselves to investigating fraud in the scientific literature. He contacted them and they in turn contacted Margot O'Toole.

With repeated appeals, Mr. Stewart and Dr. Feder convinced Margot O'Toole to give them a copy of the 17 pages of notes from Dr. Imanishi-Kari's experiments. "She had told us she was reluctant to give us the 17 pages, and she gave as one reason the fact that her husband had a job working for Wortis at Tufts," said Stewart.

Stewart and Feder then turned the information in these notes into an article of their own, titled, "Original Data Contradicts Published Claims: Analysis of a Recent Paper."[53] Their supervisors at the NIH, however, refused to give them permission to submit their paper for publication, or to contact the *Cell* paper's authors. Three months of meetings with NIH officials followed. Part of the delay was due to the process of sending their manuscript out for anonymous peer review. Eventually, they received a December 12, 1986, memo from a high-level NIH official, Dr. J. Edward Rall, who said he was "withholding approval" of the manuscript for publication "pending your attempt to find a resolution with the authors of the *Cell* paper." He included the peer reviewers' comments and wrote, "The three attached reviews of your paper are unanimous in suggesting that it is important to contact the authors of the *Cell* paper you are challenging on the basis of 17 pages of original data that you have in your possession."

Stewart and Feder attempted, for the next three months, to resolve the issue with the *Cell* authors. Their testimony to the Dingell subcommittee describes their efforts. The portion of their records devoted to directly contacting the authors of the *Cell* paper by phone or letter covers five single-spaced pages and includes 23 chronological entries from December 18, 1986, until March 26, 1987. Some of the entries are quite striking:

> *5 January 1987.* Walter Stewart phoned Dr. Baltimore, who referred to Dr. O'Toole as a "disgruntled postdoc" and said

that she was engaged in a personal vendetta. He said that what we were doing was dangerous and insulting—a Pandora's box. He stated that we should be worried about the quality of our information. He stated that he was angry and that he would look into our motives for inquiring about the accuracy of his paper. He said that if we wanted him to carry the matter further, he would, and in that case it would be his neck against ours. After further discussion, he terminated the phone call abruptly.

21 January 1987. Dr. Baltimore sent us a letter. . . . He stated that he did not "accept" the idea of our doing an internal audit. He added: "In my mind, this is a dead issue and I am encouraging my colleagues to so consider it with this letter." He sent copies of the letter to Dr. Imanishi-Kari, Dr. Weaver, and Dr. Constantini. On the same day, Dr. Wortis in a phone call said that because he had looked into the accuracy of the *Cell* paper, a reexamination by us would serve no useful purpose. After some discussion, he terminated the call abruptly.

11 March 1987. We received a letter (dated 2 March) from Dr. Wortis that said in part: "No doubt you are curious about the relationship between the information you have been given and the published material. But there is no social or scientific gain in satisfying your curiosity." The letter concluded with the statement: "Please let this be an end to our correspondence."

17 March 1987. Walter Stewart phoned Dr. Baltimore, who refused to give us access to data and criticized our motives and those of Dr. O'Toole. He characterized her objections as "nonsense," and our actions as "grossly unethical." He stated that Dr. O'Toole had stolen the documents from the laboratory under unethical conditions and that she wished to make trouble. He stated that he had seen all the data that we had seen. He stated that, in similar circumstances, any of his other colleagues would simply phone him up, inform him that they had data that had been xeroxed, apologize to him, and return the data to him. He stated that he would counter our requests in court if he had to, and that he would sue us for harassment and libel. He stated that all the assertions of fact in our draft manuscript were untrue, but he refused to give any specific examples. He stated that he and Dr. Imanishi-Kari agreed that they would discuss the question of the paper with an immunologist but not with us. He stated that his only recourse was to make us look like fools, and asked why we placed him in that position. He stated that we had written a bad faith article based on data that were stolen. He stated that our action in writing was immoral and that we would all find out whether or not it was illegal.

24 March 1987. . . . Dr. Baltimore sent us a letter in which he said that attempts such as ours to analyze a published paper would tie up the scientific community in "futile, time-wasting debate." He referred to Stewart's previous letter and said that the attempts to quote the phone conversation with him were "in extremely poor taste" and "selective."[54]

Baltimore wrote in his letter to the scientific community about the case that the coauthors of the *Cell* article had discussed whether to give Stewart and Feder the data and ultimately decided not to. Among the reasons he gave was that the two were "self-appointed and had no right to the data. We believed then, and still believe, that for random people, scientists or not, to investigate scientific papers would severely disrupt ongoing scientific activities. On the other hand, duly constituted investigative bodies or colleagues in the same field should be provided the data for investigative purposes without question, in our opinion."

Stewart and Feder never (until 1991, after Baltimore withdrew the *Cell* paper) were able to get their paper published in the scientific press. Benjamin Lewin, editor of *Cell,* turned it down, writing that he couldn't send it out for peer review because the underlying data was only 17 photocopied pages out of many others in a laboratory notebook and therefore constituted "an incomplete set" of the pages from the notebook.

He then added, referring to an earlier conversation, that "we are prepared to entertain correspondence" on controversies concerning data, but in general such letters must occupy less than two pages of the journal. Lewin didn't feel that his journal was the proper forum for "extensive analysis of fraud," but suggested instead that the NIH set up a review committee to look into the original paper.[55]

When Stewart proposed a letter to the editor, Lewin rejected this too.[56]

Stewart fared even worse with *Science.* The managing editor wrote back in a short letter that their paper could not be peer reviewed because it was "not" a scientific paper. This rejection letter too suggested that this topic was "best handled" by some sort of investigating committee.[57]

In the end, Congressman Dingell published their paper in the record of his April 12, 1988, hearing. In effect, Congress became the science publisher of last resort.

BALTIMORE STRIKES BACK AT DINGELL'S MAY 4, 1989, HEARING

"So, then, what is the case against us? . . . That we committed fraud, created data, or misrepresented the facts? This, I reject categorically. The subcommittee can point to nothing that supports these allegations," testified Nobel Prize winner David Baltimore to John Dingell's Oversight and Investigations Subcommittee (OSI) at a second [May 4, 1989] hearing.[58] Baltimore righteously defended the *Cell* paper he had coauthored with five others, including Dr. Thereza Imanishi-Kari.

Congressman Dingell, who had championed many whistleblowers in the past, had taken sides with Margot O'Toole, to the extent of bringing in the forensics laboratory of the Secret Service to determine the authenticity of notes from Imanishi-Kari's laboratory. He insisted on the intrusion of the Secret Service, which had the United States' best forensics laboratory, into the case because Baltimore and his MIT colleagues were world-renowned scientists. They could have easily dismissed the conclusions of a forensics lab that was less than the best. However, the involvement of the Secret Service was enough to send shudders through the scientific community. *Chemical and Engineering News* wrote, "There is particular revulsion over the calling in of Secret Service forensics experts. . . ."[59]

In fact, the Secret Service agents contrasted almost comically with the polished researchers from MIT. "I received a BS degree at Bob Jones University in South Carolina," said Secret Service forensics examiner Steve Herzog.

The Secret Service was reluctant to get involved, doubting their own competence in investigating the scientists' research. "At first they said they couldn't do the analysis," said Walter Stewart. Nevertheless, by the day of Dingell's hearing they had found that 19 percent of the key lab notes made by Dr. Imanishi-Kari supporting her disputed paper were suspect. Five to ten pages of notes contained dates that had been altered, apparently with the intent to conceal according to the Secret Service, and five pages of notes dated 1984 and 1985 had, according to the forensic evidence, been written up some time around May 1986. In addition, numerous pages had been prepared out of chronological sequence—1985 experiments had been written up before those of 1984. "But," acknowledged Stewart, "none of this conclusively proved that the actual experiments hadn't been done."

In her written testimony, Dr. Imanishi-Kari gave her response to this evidence by writing, "None of the information presented by the Secret Service concerning the recording of data after the date of the experiment surprises or concerns me. I freely acknowledge that in some cases I recorded lab notes after the conclusion of the experiments, sometimes probably several months after." In her oral testimony, Imanishi-Kari added, "Now I have a chance to see some of these new charges and to me they make no sense. What they seem to be saying is that I am not a neat person. Well, that is true. I do keep my notes in what seems to others as a mess condition. But I know my notes. I know where they are and how to read them, and that's what's important."

Dr. Imanishi-Kari clearly won the day with the press by testifying in her broken English. "Mr. Chairman, I have lupus. My sister died of lupus. That was in my mind all the time I was doing my research. I have hoped all along that I could help provide insight that might lead to a cure for this disease. If had fabricated data, it would have misled scientists, wasted their precious resources and retarded their efforts to cure the disease that killed my sister and threatens me."[60]

If this heartrendering testimony wasn't enough, David Baltimore totally supported Dr. Imanishi-Kari. This was particularly curious because only a few days earlier, he had seen the disputed notebooks for himself at a meeting with Dingell's staff. Peter Stockton recalled, "In 1989, when he saw those notebooks for the first time, he looked like he was going to vomit on the floor. Walter [Stewart] showed him that this could not possibly be a credible experiment." The OSI draft report, which branded key parts of the data that allegedly supported the paper as "fabricated" and led to the paper's withdrawal by Baltimore, says that Dr. Baltimore was "aware of Dr. Imanishi-Kari's having organized the notebooks to respond to the NIH and Congressman Dingell's subcommittee."

The report refers to a meeting between Dr. Imanishi-Kari, Dr. Baltimore and his attorney, Normand Smith, Jr., shortly before Dingell's subcommittee obtained the notebooks. Smith told the OSI, "Thereza came with all of her data and there was a discussion . . . as to whether we should just dump it on the doorstep of the committee . . . or should she go through her data, catalog it and put it in order and try to make it as comprehensible as possible." Although some of the data was in folders and spiral notebooks, Smith told the OSI, "there were [*sic*] a lot of just loose paper." Dr. Baltimore told the OSI that Dr. Imanishi Kari went home with the data to organize it ". . . entirely on her own . . . over a weekend."[61]

Despite this, Baltimore testified on behalf of Imanishi-Kari: "She is also a victim. . . . She deserves my support, and the support of all scientists, for any of them could be in her shoes."[62] As for Dingell's use of the Secret Service, Baltimore referred to the horrified reactions of other scientists when they heard about it: "Recently they have read stories about this Subcommittee calling in the Secret Service to analyze lab notes and we've now seen the results of that. Mr. Chairman, you must agree that their [the scientists'] concern is understandable." He himself was concerned, "I must tell you, Mr. Chairman, I am very troubled about how this situation has got so out of hand. I have a very real concern that American science can easily become the victim of this kind of government inquiry." Baltimore then brushed aside what the government investigation had found. "There is still nothing from the Secret Service investigation that causes me to doubt the validity of the *Cell* paper," he testified. He even made the Secret Service look foolish: "They discovered that Figure 4 in the paper was a composite of different exposures of a single autoradiogram. I would have told them that. This is a common practice in presenting such data."

Baltimore made more media points in a final, dramatic showdown with Dingell that dealt with several different issues. Concerning one of them, Baltimore said, ". . . a member of your staff compared scientific fraud of which I am accused . . . [to] the Nazi Holocaust. That's a quote from Mr. Walter Stewart, and it was quoted in *Science Magazine* in February, 1989 . . . it's reported that Mr. Stewart came to the front of the room and wrote on the blackboard the word 'Holocaust' and I quote from *Science Magazine*, 'By this, Stewart means that the problems of scientific cheating were being ignored by many researchers who, like some Germans, dealt with the problem by looking the other way.' It was not the best analogy tc use before an audience of scientists where more than a few are Jewish. So, my charges, sir, are not made lightly."

Supporting Dr. Baltimore, packing the subcommittee room, sat angry scientists, outraged with Dingell's intervention into their turf, and alarmed at his use of the Secret Service.

THE RESPONSE OF THE SCIENTIFIC COMMUNITY

The scientists were present at the second subcommittee meeting because, about a year after Baltimore wrote to the scientific community, MIT's Dr. Philip Sharp sent out his own "Dear Colleague" letter, on April 18, 1989, a few weeks before Baltimore's congressional appearance. Sharp not only asked for scientists to send letters to their congressmen and to write op-ed pieces, he included a sample of such a letter and a sheet listing "points that can be made." *Science & Government Report* said: "Sharp told SGR that members of Baltimore's staff at the Whitehead Institute helped him prepare the material. In various forms, the stuff, highly selective in content, showed up in many major publications, usually on the theme of bullying Congressman assaulting a respected leader of science."[63]

One of the sentences in Sharp's sample letter states that Dingell's investigation "has caused serious confusion about the difference between error and fraud in scientific research."[64] Robert E. Pollack, the dean of Columbia College and a professor of biological sciences, wrote a *New York Times* op-ed piece titled, "In Science, Error Isn't Fraud." He concluded, "I fear the way Dr. Baltimore is being treated means that witch-hunts are in the offing."

Others, compared the problems David Baltimore was having with Dingell to those Galileo had with the Medicis. Stephen Jay Gould, professor of geology at Harvard, wrote in *The New York Times:* "We continue to deplore Galileo's fate and rank him first in the noble army of scientific martyrs. And yet, in the light of recent developments in Washington, I'm not so sure that Galileo might not be in more trouble today. Several Congressional committees have been investigating scientific misconduct and some seem ready to view error as a cause for investigation into the misuse of Federal funds. On this model the Medicis of Florence might consider prosecuting Galileo for his misreading of Saturn."[65]

Dingell's hometown paper, *The Detroit News,* also used the Galileo theme in an editorial titled, "Dingell's New Galileo Trial." The paper stated, "Congress ruling on the accuracy of complex scientific experiments makes about as much sense as the clergymen who tried Galileo over his astronomical observations."

An echo of the Galileo theme appeared in *The Washington Times'* editorial of May 11, 1989, titled "Dr. Dingell's Weird Science."

Paul Gigot of *The Wall Street Journal,* referring to Dingell as "Congress's grand inquisitor," then quoted David Baltimore, who compared Dingell to Joe McCarthy: "'It has been a revelation to me the kind of power that a man like John Dingell has,' says Dr. David Baltimore, a rational man suddenly exposed to the rituals of Washington. 'I had no idea . . . Congress has a reputation that I thought had been cleaned up from the days of McCarthy and McCarran. It's clear when you look at Dingell that that's not true.'" Gigot concluded, "The arrogance bred of unchallenged power has stripped them of self-restraint and distorted their understanding of the public good."[66]

The Wall Street Journal's own editorial writers warned, "David Baltimore's travail is only the beginning if scientists remain silent and let John Dingell become the Auditor General of American science."

The lobbying campaign was effective, but also expensive. The Whitehead Institute at MIT, which had hired lawyers from various Washington law firms to lobby Congress and advise Baltimore and others, then claimed this financial outlay as overhead expenses on its government grants and received reimbursement for it. The Institute later agreed to give back $69,000 to the government, and MIT withdrew $27,300 it had billed to the government on this issue.[67]

Asked in a telephone conversation what he thought had been the most extraordinary aspect of this case, Walter Stewart replied, "The overwhelming reaction of the scientific community to demand an end to the investigation without even examining the evidence or seeming to care what happened to the one totally and obviously blameless person, Margot O'Toole."

BALTIMORE AND THE NIH REPORTS

At the May 4, 1989, Dingell hearing, Baltimore testified that the subcommittee "have done the most extensive analysis of scientific data probably the world's ever seen.

"And what they found out is that all the data in the *Cell* article are not in question; that the only questions that have been raised, and I think they're artificial questions, but the only ones that have been raised deal with pages that did not involve data in the *Cell* paper."

There was a bit more to this than Dr. Baltimore suggests. The data in the photocopied 17 pages definitely did not support the conclusions in the paper. The NIH panel appointed in 1988 to look into the matter established this at once. However, Imanishi-Kari produced a lab book purporting to describe data from other experiments conducted at about the same time that did support the conclusions, specifically the data presented in the major table of the paper, Table 2. An exchange between Bruce F. Chafin, a special assistant to the subcommittee, and NIH panelist Joseph M. Davie, Searle's president for research and development, explains what happened. Chafin asks, "In the *Cell* article, that experiment that was reported to be done was not, in fact done?"

Davie replied: "Was not done. . . . Instead there were other experiments that were not described in Table 2, in which that kind of experiment, in which that very experiment had been done and the results in that experiment were that, in fact, the particular class of immunoglobulin was found. The mistake was in saying that they had done that experiment in the animals that were described in Table 2. They did not do that. They, in fact, had done it with different animals and had come to that conclusion."

So Baltimore is correct when he suggests the forensic analysis "did not involve data in the *Cell* paper." The NIH panelists saw the data in question to be so obviously bad that they were not even subjected to forensics analysis. As a result of the lab book submitted by Dr. Imanishi-Kari, the NIH panel called for a letter to *Cell* informing the readers about the "true" source of the data for the key Table 2. Margot O'Toole explained, "The panel finds that I was correct about these data but describes subcloning experiments that do support the Table 2 claims and recommends that these be published instead."

Dingell brought in the forensics experts because O'Toole had evidence that the notebooks containing these "other experiments" had been faked. O'Toole testified, "[A]t the time of my initial challenge, I asked Dr. Imanishi-Kari if these subcloning experiments had been done. If the experiments had been done and supportive results obtained, I would have withdrawn my challenge. Dr. Imanishi-Kari stated that the experiments were not done. . . . In November 1988, I told the NIH that the panel had relied on data that I was told did not exist at the time of the challenge and they disregarded what I said.[68]

Dingell was curious about how much data the panel had actually looked at. He asked Dr. Davie of Searle, "Dr. Davie, you testified that the panel reviewed all the data published in the paper. Is that correct?"

Davie replied "Yes."

But then after an exchange on some "data" that didn't exist, Dingell asked, "So how did you review then the data here that does not exist?"

Davie replied, "Again, we did see the data that was in our initial. . . ."

Dingell cut him off, "I think we can simplify this by pointing out that your statement then—that you had reviewed all the data—was incorrect."

Davie answered, "Well, that's correct. That's correct."[69]

O'Toole summarized the significance of this omission. "The evidence uncovered by the Secret Service calls much of the unpublished data into question. The authors reply that none of the questioned data was published. What is crucial to remember is that although the questioned data were not published, the panel relied on them in deciding there was experimental evidence in support of the central claim."[70]

The NIH investigation concluding that the main "data" were faked was the *second* NIH investigation into the case. It was started in early 1989, at about the same time that Dingell's Secret Service investigators from Northern Virginia Community College had raised serious questions about the MIT data. Was the Secret Service investigation the reason for the second NIH investigation? Not according to NIH Director Wyngaarden: "Dr. O'Toole requested a copy of the data for Table 1, and the staff contacted Dr. Imanishi-Kari to ask her where all of those data points were and then sent the relevant part of the record, a copy of it, to Dr. O'Toole.

When she examined it, she found that data for some of the points on the figure were missing and asked about that. We then contacted Dr. Imanishi-Kari, who contacted her colleague, Dr. Reis, in Brazil, who wrote a letter explaining that that experiment had been done in two parts. The highest dilutions had been done the following day and the data entered directly on the chart. . . .

"Now that may be a perfectly logical explanation, but it did raise in our mind the question of whether there were other specific data points that also were not represented in the record, and we decided that Dr. O'Toole's question had merit and we should take it seriously. On that basis we opened the investigation."[71]

BALTIMORE'S LETTER AND CONFLICTS OF INTEREST OF THE NIH PANELISTS

The composition of the 1988 NIH panel is itself quite interesting, particularly because Baltimore himself had called for the formation of an NIH investigatory panel.

After the NIH appointed the panel, Congressman Dingell said to top NIH officials: "I find it curious that on the panel that NIH and the HHS selected to review the behavior of Drs. Imanishi-Kari and Baltimore, you have three individuals, one of whom is a coauthor of a textbook with Baltimore and another who has coauthored 14 papers with Baltimore."

Dingell asked Robert Windom, the Assistant Secretary for Health, the Department of Health and Human Services, "Now sitting down there objectively, would you tell me that they would give you a fair evaluation?"

"It would be difficult to say so," replied Windom.

The NIH replaced the two questionable panelists and the panel issued its report, calling for a correction to be published in *Cell*. As Wyngaarden wrote in a January 31, 1989, letter to Baltimore, "no evidence was found of fraud, misconduct, manipulation of data, or serious conceptual errors." Wyngaarden did give Baltimore several slaps on the wrist. He requested that the "letter or manuscript of correction which you and your colleagues would submit to *Cell* in response to this decision should be sent to me *prior* to your sending it to the journal, so that we may review it for completeness and accuracy." Nobel prize winners usually do not receive demands such as that. The second slap had more sting: "It appears that even though the allegations have been known to you and the other coauthors of the *Cell* paper at least since the Spring of 1986, the coauthors never met to consider seriously the allegations or to reexamine the data to determine whether there might be some basis for the allegation. Such an analysis on the part of the paper's coauthors, followed by appropriate action to correct such errors of oversights, may well have made a full investigation unnecessary."

THE EXTRAORDINARY TURNAROUND

Nearly two years later, in March, 1991, after a second Secret Service investigation and another, May 1990, hearing by Dingell's subcommittee, at which the Secret Service reported much of its findings, the facts trumped the responses of the angry scientists who had righteously supported Baltimore. The NIH's Office of Scientific Integrity concluded: "The forensic evidence and the extensive statistical analyses establish that the June subcloning data and the January fusion data are fabricated. It remains unclear if these experiments actually were done . . . But if they were, the resulting data must have been deemed unreliable or otherwise unsatisfactory, since an attempt was made to substitute other data for them."[72] Dr. Imanishi-Kari continued to insist on her innocence. She was quoted as saying, "I know it sounds tacky, but I am innocent. I really refute all of this stuff."[73]

The OSI criticized Baltimore for his flat-footed defense of Imanishi-Kari. "[I]t is difficult to comprehend his maintaining this stance as the evidence mounted that serious problems existed with the seriological data in the *Cell* paper." The Report made withering comments on Baltimore's testimony to the OSI committee: "Dr. Baltimore's most recently-expressed views concerning the investigation are the most deeply troubling. These were statements Dr. Baltimore made on April 30, 1990, when he was interviewed by the OSI investigative team. Dr. Baltimore disputed the significance of the June subcloning data and he asserted that if they were fabricated, the NIH was somehow responsible for this act of scientific misconduct: 'If those data were not real, then she (Dr. Imanishi-Kari) was driven by the process of investigation into an unseemly act. But it does not go to the heart of any scientific issue. . . .'

"Dr. Baltimore went on to say, '. . . if something is not published, it's in your notebooks and it's not published, that it is not then a matter for those rules to be followed. . . . [I]n my mind you can make up anything that you want in your notebooks, but you can't call it fraud if it wasn't published. Now you managed to trick us into publishing—sort of tricked Thereza—into publishing a few numbers and now you're going to go back and see if you can produce that as fraud. But I think you should see that as a forced situation . . .'

"The OSI found Dr. Baltimore's statements to be extraordinary."[74]

At his subcommittee hearing of May 9, 1989, Dingell said of Margot O'Toole, "The visible result of all this is that Dr. O'Toole is sitting in this room with her career in shambles and a number of other people are prospering mightily. . . . I'm curious, where is the justice in that situation?"[75] Some of it appeared in the March 1991 OSI draft report, which stated, "Dr. O'Toole's actions were heroic in many respects. She deserves the approbation and gratitude of the scientific community for her courage and her dedication to the belief that truth in science matters."

After the OSI draft report was leaked to the press, Baltimore announced he would request the journal to "retract the paper."[76] On May 3, 1991, he wrote the NIH, "I wish to state that if Dr. Imanishi-Kari did falsify data or make misrepresentations, I had no knowledge of the misconduct."[77]

About eight months earlier, Dr. Margot O'Toole had found her way back into science, doing basic research at a private biotechnology company.

CONCLUSION

Although there is no way to know how prevalent data faking such as that discussed in the Baltimore case may be, the appalling reactions of the scientific institutions and community are essentially consistent in all three cases discussed in this chapter as well as in the cases discussed in Chapter 6. The fundamental reason for this consistency, the nature of the science patronage system itself, will be taken up in Chapter 8.

Margot O'Toole, in her comments on the 1991 OSI draft report, stated a fundamental conclusion of the Baltimore case: "It surprises me that the principle which guided my actions is not a universally accepted one among scientists. For instance . . . an eminent scientist who has been, and continues to be, a most outspoken critic of what I did . . . was invited to deliver an important public seminar on scientific misconduct last month.

"This shows that this scientist is regarded, at least by some, as an expert on the ethical issues facing scientists.

"He and others have stated that, in this case, the protection of careers must take precedence over scientific accuracy. This closely echoes what Drs. Wortis and Huber said to me in May 1986, what Dr. Baltimore wrote to Dr. Eisen in September 1986 and what the reviewers of the Stewart and Feder manuscript said. . . . I therefore sadly conclude that the attitude that scientific careers are much more important than science has become common among scientists."[78]

CHAPTER 5

IMPURE PROFIT: MARKETING MEDICINE IN SCIENTIFIC RESEARCH

Because medical research conducted exclusively by or for major pharmaceutical corporations is an area of science that involves life and death issues, government regulation is necessarily heavy. The workings of the giant pharmaceutical companies usually remain hidden behind corporate walls, but when something goes wrong and people die unnecessarily, disturbing documentation occasionally comes to light. According to a major congressional report, such documents reveal that violations of Food and Drug Administration (FDA) regulations "may have reached epidemic proportions."[1]

No one case can fully expose the range of abuses that have become all too common in the drug industry. Therefore, I have woven together several cases, focusing most heavily on three: the development of a pain killer known as *Zomax**; the *Bjork-Shiley convexo-concave heart valve,* and the anaesthetic *Versed,* used in surgical procedures requiring conscious sedation. The evidence shows that corporations often jettison sound judgment and sound

* Zomax is a registered trademark of McNeilab, Inc.

methods in the pursuit of profits. The market for the class of drugs to which Zomax belongs, for example, is worth billions of dollars. "The companies try to shove something onto this market, get people using it, because there is a strong brand loyalty," said Allan Kanner, a well-known plaintiff's attorney who is handling the key Zomax case involving corporate insiders.

Some abuses arise because an inherent conflict of interest undermines the use of the scientific method. For example, often the same person charged by the company with responsibility for gathering information on adverse reactions to a drug is also charged with selling and promoting the drug to the doctors.

A second phenomenon undercutting the integrity of the medical science in such cases is *concurrency,* the simultaneous research and production of a new drug. This notion occurs most commonly in Pentagon-sponsored science, usually to disastrous effect.

A third phenomenon interfering with the corporate scientific process is secrecy. In itself, secret corporate science is hardly surprising and may even be desirable to protect valid trade interests. But too often, companies take the need for secrecy much further, requesting, for example, protective court orders to prohibit the gathering and distribution of critical data of life-or-death importance to the potential users of their products. Manufacturers have even intervened to discourage the publication of scientific articles that would have warned practitioners and other researchers about such dangers. These actions defeat the free flow of ideas on which the advance of science depends, prevent the dissemination of information on potential dangers resulting from the science, and interfere with the machinery of government regulation in situations where the scientists are clearly not regulating themselves.

DRUG-INDUCED ANAPHYLACTOID SHOCK

Allergic Reactions to Zomax

"A lot of physicians saw me and thought I was clinically dead because I didn't have a blood pressure, I was blue, and it looked like I had brain death," testified Glen Stanbough, MD, and professor at a Texas medical school. He was speaking under oath about the life-threatening anaphylactoid reaction he suffered on January 16, 1982, after he took a McNeilab, Inc., painkiller, Zomax.

Anaphylactoid shock is an extreme allergic reaction in which the body literally shuts itself down; within a few minutes the heart slows or stops altogether, the throat and tongue swell to choke off breathing, and the victim loses consciousness. It is often fatal.

Devra Davis, a Zomax survivor, and a senior member of the American College of Toxicology, testified, "[A] person can die from exposure to Zomax, without a hint that they are at risk." She added, "Zomax is one of the few compounds ever identified that can spawn powerful immunological reactions in people with no previous history of allergic response."[2]

Dr. Stanbough testified that when he tried to find out more about Zomax from McNeil, the answers he received were not helpful. "I was visited," he said, "by the medical director of McNeil Laboratories . . . she told me that I was an unusual case, that it was an isolated phenomenon in Lubbock, and that they had no previous reports of reactions that severe."[3] Stanbough testified that the McNeil medical director visited him on April 19, 1982. However, her reported claim that his reaction was unusual is contradicted by the congressional testimony of the FDA's Dr. Harry Meyer. "You had 25 cases by March of 1982 and you had your first death from anaphylaxis. . . ." In a February 5, 1990, sworn deposition taken by Detroit plaintiff's attorney Carol Youngblood, the medical director stated that she could not recall any of the names of the persons she spoke to or medical facilities she had visited in Lubbock during the period in question. However, she stated that she did "not necessarily" conclude at that time that the strong allergic reactions in Lubbock were caused by Zomax, in part because the details of some cases were "impossible" to track down. As for those she could track down, she did not remember her conclusions. In her deposition, the only death she referred to for this period in early 1982 was that of a "known aspirin-sensitive person." The medical director stated "it was never proved whether or not he ever took the drug [Zomax] or not."

Anaphylactic and anaphylactoid reactions involve essentially identical symptoms even though anaphylactic reactions have an immunological base and anaphylactoid reactions do not. "The terms do not distinguish severity, but biochemistry," explained Ed Lemanowicz, PhD, who during most of the events covered in this case dealt with regulatory (FDA) affairs for McNeilab, Inc. In the reported Zomax cases the reactions typically followed the same pattern. Most of those convulsed by the drug had taken Zomax previously with no reaction; had stopped taking it for a period of days, weeks, or months; then had restarted it. Within 20 minutes of swallowing a Zomax, the reaction begins.

Stephen Alexander, MD, summarized the reaction in three of his patients: "All of them described an itching sensation which began in the lower extremities or in the groin, in the vaginal area, which then spread upward into their abdomen and then chest, thorax, and face.

"Two . . . described a very raspy voice. They could hardly recognize their voices as being their own. They had a thick tongue. They became short of breath. They had difficulty breathing, and they noticed that their pulse was racing and that they were profusely sweating.

"All three patients were so frightened and disoriented that they were confused, didn't know what to do, and . . . were rushed to the emergency room."[4] Frequently the victims pass out, usually within 45 to 60 seconds of the onset of the reaction. "Every time I pick up a paper now and read of a . . . crash of a private plane with an unexplained reason my first thought is: Did he take a Zomax?" said Jack Yoffa, another physician.[5]

The reaction can be fatal if not treated in time or properly. Unless emergency room staff know what to look for, they may not give appropriate treatment, which usually involves injecting the victim with Adrenalin and often with an antihistamine. Unfortunately, if the victim has a history of heart disease and the emergency room staff doesn't know it, Adrenalin itself can be lethal.

"In August and September 1982 . . . Zomax . . . was described as being safer than aspirin," said Dr. Alexander, "It took a period of time for all of my three patients to be treated adequately simply because that kind of reaction was unknown."[6]

Zomax, this potentially fatal drug, is a prescription, nonsteroidal anti-inflammatory drug (NSAID). Aspirin is the best-known NSAID. Zomax was given FDA approval October 28, 1980, for patients experiencing mild to moderately severe pain. McNeil Pharmaceutical, an operating division of NcNeilab, Inc., itself a wholly owned subsidiary of Johnson & Johnson, developed and marketed the drug, giving it to 3,600 patients during clinical trials in the United States, starting in 1974. It produced tumors, diagnosed as "malignant,"[7] in rats during laboratory tests and was thus assumed to have the potential of the same effect on humans. In fact, the doses given to the rats were proportionately much *smaller* than human doses would be, a fact described by one FDA expert as "scary."[8] The package insert for Zomax specifically recommended against long-term use or for use on children. Despite this warning, about 20% of Zomax users took it chronically, day after day, month after month.

Tolectin—Zomax's Shocking Twin

Zomax differs by only one molecule from another McNeil NSAID, Tolectin. Because two drugs are nearly identical in structure, however, does not necessarily mean that they are associated with the same reactions in the human body. Dr. Arthur Hull Hayes, Jr., Commissioner of the FDA, testified: "Sometimes very small differences in what are otherwise . . . almost identical substances, can be extremely different. . . . Indeed, in the family of drugs in which aspirin is one, the salicylates, the smallest changes in the molecule change the pharmacology and the toxicity of those drugs enormously. Sodium salicylate . . . is a pain reliever. A very small change in that compound, in fact, makes it a keratolytic agent that is quite toxic but is great for dissolving warts. . . . Alternatively, the two drugs may turn out to be very much alike. . . . [B]ecause they are chemically similar and have pharmacologic similarities, then perhaps they will give the same adverse reactions."[9]

The latter is precisely what happened with Zomax and the earlier McNeil drug, Tolectin. After the FDA approved Tolectin for arthritis pain, inflammation, and swelling, the drug was discovered to be "associated with" (the word *caused* is seldom used by pharmaceutical companies) anaphylactic shock. The agency put out an ADR (adverse drug report) highlight on it on June 20, 1979, reporting 12 cases of anaphylactic reactions, 7 of which were associated with restarting the drug after not taking it for a while.[10]

By January 1981, Allen C. Rossi, DDS and other researchers in the FDA's Screening of Adverse Reactions Program had completed a paper that reviewed 25 cases of anaphylactic shock associated with Tolectin. The abstract stated: ". . . it appeared that this reaction was reported with greater frequency in tolmetin [Tolectin's generic name] users than among users of other nonsteroidal anti-inflammatory drugs to which it was compared. Further, there was a strong suggestion that individuals previously exposed to tolmetin therapy were more likely to develop an anaphylactoid response than those who had not. Both suspicions have important clinical implications and require further confirmation."[11]

Because Tolectin was by this time widely used (i.e., more than 10 million prescriptions had been filled), the paper's reported results were obviously of urgent importance. Other researchers and physicians at hospitals, nursing homes, and in private practice needed that information. They didn't find out, however, until about a year and a half later when the *New England Journal of Medicine* published a letter from Rossi and an associate on August 19, 1982, by which time McNeil knew of at least one death from the nearly identical Zomax. The original article had been rejected by another journal long before. The company denies all wrongdoing.

On January 22, 1981, Rossi et al. had submitted their article to *The Annals of Internal Medicine,* a highly regarded journal read by thousands of practicing physicians around the country. The day before, on January 21, McNeil also received a copy of the article. "It was sent by the director of my division," recalled Rossi in a telephone interview with me on March 18, 1991. "At that time we had a director who insisted it be sent out," he recalled. Rossi objected to the article being sent when it was, "If the article had been accepted, but before it appeared in print—at *that* time it would only have been fair that the company get it. I'd have had no objection; the company might have critical comments or rebuttal that would have been worthwhile. I don't really feel that the scientific method had a chance to work."

Rossi received a letter from McNeil's Director of Regulatory Affairs on March 22, saying "we have reviewed" his article. Apparently McNeil didn't like the article. According to a lawsuit originally filed by two former McNeil scientists, James A. Dale, MD, and Edward F. Lemanowicz, PhD, the company was engaged in a "conspiracy to . . . cover up" the adverse reactions to Zomax and this letter was part of that effort. Dale charged the company and its parent, Johnson & Johnson, with wrongful discharge, Lemanowicz charged the company with wrongful "constructive" discharge, meaning he alleges the situation was made so intolerable for him that he quit. They filed under the New Jersey and federal civil Racketeer Influenced and Corrupt Organization (RICO) statutes. On February 13, 1991, Federal Judge Dickinson R. Debevoise dismissed a number of Dale's key complaints on the grounds they were "time barred." According to a subsequent, August 12, 1991, opinion of the judge, "Thereafter, Dale voluntarily withdrew his remaining claims." In this opinion, the judge issued a summary judgment in favor of the company and dismissed Lemanowicz's complaints. In doing so, however, the judge did not dispute any of the documented evidence or factual allegations concerning Zomax that will be discussed in this chapter, but rather said that even if all these factual allegations were assumed to be true, Lemanowicz "was not placed in an intolerable situation that compelled him to quit his job with McNeil." Lemanowicz is appealing.

The legal brief of the two former McNeil scientists states that the McNeil letter contained "highly critical" comments. McNeil, which denies all of their allegations of wrongdoing, says the letter "speaks for itself" and specifically denies that the letter was sent "in furtherance of a fraudulent drug marketing conspiracy."[12] McNeil also sent a copy of this letter to the *Annals of Internal Medicine.* "I thought it was inappropriate that the company should have written them [*The Annals*] at that time, if they did write them," said Rossi. A physician from a major medical school who was a member of the journal's articles review board, was also on retainer as a consultant to McNeil, a fact that McNeil admits.[13] In an April 2, 1981 letter, the review board rejected the article for publication. Rossi commented: "The article was reviewed by three peer reviewers. Although I got copies of their reviews, I don't know who the reviewers were. Two of the reviewers were skeptical of the research reported in the article, as I was myself, but felt that I had made a balanced presentation. Those two reviewers recommended the article be published. The third reviewer seemed to go out of his way to poke holes in it. The points the third reviewer made at times seemed outrageous. I thought he was more of an advocate and less of a scientist."

The vote was two "yeses" and one "no," but the "no" won out. Majority rule didn't seem to apply in this instance. Rossi, who has since published in many prestigious journals, reports that he has never submitted again to the *Annals.* The secrecy of the peer review process prevents us from conclusively identifying the author of the negative review.

According to the Dale and Lemanowicz lawsuit, the medical director of McNeil told Dale that McNeil would "kill" the article. McNeil specifically denies this allegation. In a January 1992 interview with me, Lemanowicz stated that the medical director told a group of McNeil employees that the company had a way to "block the publication."

Internal McNeil documentation, however, addresses activities by the company concerning the article. On September 8, 1984, Patrick H. Seay, PhD, McNeil's Director of Regulatory Affairs, wrote a lengthy memo (22 pages), hereafter referred to as "Productivity" that he sent to R. Z. Gussin, who was then the Vice President for Scientific Affairs. Seay typed at the bottom of his memo a request that it not be copied. In this memo, Seay recounted many of the successes during his 16 years in his job. He also noted that some activities had "not enhanced our reputation." Among the examples, he cites anaphylactoid reactions to Tolectin. The memo states that the company tried to "blunt the effects" of two articles, one of which was the 1981 Rossi article.[14] In the memo, Seay writes that the company should not oppose the submission of scientific articles by FDA researchers concerning possible side effects of any individual examples of a class of drugs, provided the article is submitted to a reputable, refereed journal.

Anaphylactic Reactions during Zomax's Clinical Trials

Since Tolectin was producing relatively large numbers of cases of anaphylactic shock and Zomax was an almost identical drug, it would have been reasonable for the company to be concerned about the likelihood of similar reactions from Zomax. The clinical trials of Zomax on 3,600 patients took place between 1974 and 1980.

A January 6, 1978, internal McNeil memo, authored by Frederick L. Minn, MD, PhD (chemistry), warned the medical director that one patient suffered an extreme allergic reaction—"lightheadedness, diplopia, sweating, bronchospasm." The memo alerted the medical director "the diagnosis of idiosyncratic reaction was almost certain by virtue of our experience with the cogener tolmetin."[15] McNeil admits only that "in January 1978 Dr. Frederick Minn authored an internal memorandum which is contained in the FDA files for Zomax."[16]

However, the September 8, 1984, memo from Patrick Seay stated: "I do not believe we could reasonably have predicted the magnitude of the problem. However, we knew the chemical relationship of Zomax to Tolectin and we knew that Tolectin produced anaphylactoid/anaphylactic reactions. A member of management said that he would start to worry when we received the first anaphylactoid reaction. Dr. Minn called attention in a memo 6 January 1978 to the fact. . . ."

More warning came in a July 7, 1981, internal memo to McNeil's medical director from J. W. Gorder, MD, who was originally in charge of monitoring adverse reactions to Zomax.[17] The memo summarized a meeting of McNeil researchers and managers on anaphylactoid reactions to Zomax and Tolectin. The memo stated that the medical director "raised concern that there may be a higher correlation (cross sensitivity) between TOLECTIN allergy and ZOMAX allergy than between these two and other NSAID's." The memo added that data obtained so far from reports of adverse reactions "while scanty . . . would tend to confirm this idea."[18]

In fact, the similarity between Zomax and its McNeil precursor, Tolectin, later caused an FDA official, Dr. Robert Temple, to testify before Congress about a colleague's dealings with McNeil: "I believe . . . what he was telling McNeil then is that they were making a phony distinction, at least one he didn't believe, between Tolectin and Zomax, and that it was just a promotional gimmick, and that they should do everyone a favor and eliminate all of the discussion by getting rid of it [Zomax]."[19] Seay's September 8, 1984, memo on productivity stated that McNeil had not made a scientific discovery which resulted in a new product actually being put on the market in 17 years.[20]

Strong allergic reactions to Zomax were reported to McNeil in 1977 and 1978, and McNeil's own internal reports on them are included in the Weiss subcommittee record of the hearing. The subcommittee submitted the records of five of these cases to a Washington allergist who had previously been an immunologist at the National Cancer Institute. Two of the cases occurred in individuals with a history of allergies, so an allergic reaction might have been "expected," said the physician, Daniel Ein. As for the other three, he described them as "anaphylactoid reactions," which hit "individuals without any previous allergic history and in whom there was no reason to anticipate potentially serious reactions. One of these three individuals in fact had a life threatening reaction with respiratory impairment requiring intensive treatment in a hospital emergency room."[21] Testifying at the Weiss hearings, the group leader for FDA's Zomax review, Dr.

John Harter, said, "The third case is a more serious allergic reaction and some people would classify it as anaphylactoid."[22] Apparently, however, the McNeil scientists did not. According to a top McNeil scientist in his prepared statement to the Weiss subcommittee in April 1983, "No anaphylactic reaction had occurred in those clinical trials."[23]

In his September 8, 1984, memo, Seay reported that in the course of the clinical investigations, a "considerable number" of patients exhibited symptoms "suggestive of allergy/anaphylactoid/anaphylaxis."

In early 1985, Dr. Harter of the FDA took a thorough look at the data from the preapproval clinical trails. On February 14, 1985, six McNeil scientists met with Harter and two other FDA officials to discuss Harter's findings. Two of these scientists, James Dale, MD, and E. F. Lemanowicz, PhD, then sent a February 19, 1985, internal memo to a Johnson & Johnson lawyer. The two wrote that Harter had gone over the official filings on the preapproval clinical trials and had found 23 examples of allergy/anaphylaxis. In addition, Harter's review found that reactions to Zomax occurred between 3 and 7 times more often than reactions found simply to aspirin in the clinical trials.[24] McNeil now admits "that Dr. Harter identified instances where symptoms could conceivably be characterized as anaphylactoid reactions."[25]

According to Lemanowicz, "Harter said at that meeting, 'We (at the FDA) have to know how to deal with this and so will you if Zomax gets back on the market.'"

Perhaps these meetings had some influence on McNeil's 1985 decision against the reintroduction of Zomax.[26]

What McNeil Knew and What They Told Doctors in Lubbock, Texas

A March 12, 1982, internal FDA memo described its investigation into four Zomax cases in the Lubbock, Texas, area. A professor at the Texas Tech University Medical School "had informed the Medical Services Staff at McNeil . . . that the Medical School Staff had observed 4 significant reactions to Zomax within a 20 day period. [The Professor] received a letter from McNeil Pharmaceuticals dated January 22, 1982."[27]

The letter was signed by a physician who was McNeil's Associate Director of Medical Research and Services. It stated that after going over their files on anaphylactic and allergic reactions there weren't enough cases to even approximately estimate an interaction between Zomax and patients taking beta-blockers. The letter noted, however, that the adverse reaction reports from practicing physicians were often "incomplete" and that an interaction had not been suspected in the past.

McNeil's documented response to the Lubbock medical school professor is extremely interesting in that, only seven months after the FDA had approved Zomax, it "was already third among the drugs in its class in numbers of reports of anaphylactoid reactions associated with its use," according to Congressman Weiss. The FDA had issued a May 26, 1981, report, called an Adverse Reaction Highlight (ADR) on these reactions for this kind of drug. Although Zomax, with 14 reactions, ranked third in total number of reported anaphylactic reactions, in fact it ranked first per one million prescriptions written. The ratio for Zomax was 5.5 per million prescriptions. Ranking second per million prescriptions was another McNeil drug, Tolectin (43 reactions), which differed by only one molecule from Zomax and scored 4.30 per million. Both of these nearly identical drugs were scoring far higher than the third ranking one, Fenoprofen (7 reactions), which scored only 0.55 reactions per million prescriptions. By comparison, a well known and widely used NSAID, ibuprofen, had 0.21 reactions per million prescriptions.[28] According to Dr. Harry Meyer of the FDA, this sort of analysis by prescription, although not "true incidence data" can "raise your index of suspicion and cause you to seek other methods of refining that index of suspicion."

On July 23, 1981, McNeil changed Zomax's package insert. In the section dealing with precautions, they added a mild warning, "As with other nonsteroidal anti-inflammatory drugs, anaphylactoid reactions have been reported."

The March 12, 1982 internal FDA memo also noted, "The Medical School Staff had heard that the Medical Director and a sales representative of McNeil Pharmaceuticals had been in town after the report of the episode to McNeil and were upset because neither individual contacted anyone on the Medical School Staff during the period of the visit to Lubbock, Texas."

The medical director testified eight years later about this visit. "I spent most of my time with the sales rep from that area. . . . basically because I wasn't familiar with the area and so he accompanied me to most places. Whether he actually went into the meetings with me, I don't remember, but he certainly accompanied me on most of my evaluation."[29] She testified that she went to Lubbock in this connection twice, "The first time I went, in follow-up with Zomax adverse reactions, I saw several people. I don't remember all the names, but I saw several people." Asked if she could recall the names of any of the people she spoke to, she replied, "I don't remember." Asked about her second visit, she testified, "I think I saw basically some of the same people I had seen before. I don't remember any of the names."

The Lubbock doctors whom the McNeil medical director hadn't seen had good reason to wonder why she hadn't dropped by. Two months earlier, in January, the County Medical Society for Lubbock used its bulletin to warn local medical doctors to stop prescribing Zomax until either the FDA or McNeil clarified the association between the drug and anaphylaxis. Of the seven hospitals in Lubbock, six pulled Zomax from their formularies.

During one of these visits, on April 19, 1982, the McNeil medical director visited Dr. Stanbough and reportedly told him that his case was unusual. Asked eight years later in a sworn deposition, "Did you come to the conclusion that Zomax played a part in those [Lubbock] reactions," the director testified: "I came to no conclusions as to why there was a particular cluster of side effects in that area. Just didn't see any apparent reason. . . . I don't remember the specifics of those reactions. . . . I don't remember the specifics of those cases, I really can't tell you what the conclusions were."

What McNeil Told Doctors in Michigan and Ohio

In November 1981, a patient of Kenneth Berneis, MD, a country doctor in Plainswell, Michigan, suffered a severe reaction after taking Zomax. The man "was rushed to the emergency room with the closing off of the throat and inability to breathe."[30] Dr. Berneis added, "This first case was reported to a McNeil man who visits our office. It was implied that they had never heard of this, that it must be some sort of a weird allergic reaction."

The next month, December, another man reacted to Zomax. According to Dr. Berneis, "His wife found him on the floor totally unconscious and red. Then he became purple and was not breathing. . . . the ambulance . . . rushed him to the emergency room where . . . we could get no blood pressure. We finally got one mechanically with an ultra sound machine. . . . He had normal sinus rhythm, but we thought he was having a heart attack. He had all the symptoms. He was cold and blue and wet. . . ."

This man had apparently taken the drug by mistake. "[W]hen he had lived in Ohio . . . he had been given some Zomax and he had never made it out of the hospital door. He collapsed in their hallway, and he was told never to take it again. However, apparently he hadn't thrown it all out, and . . . mistook it for [another medicine]."

Dr. Berneis, who then reported this second case to a McNeil representative, testified: ". . . [We] talked with the McNeil representative when he came around. . . . He implied that he never had heard of this and that we really didn't want to report it to the FDA because I would end up with a lot of paperwork and I didn't have time for that."

What did the McNeil man advise the doctor to do? ". . . [W]rite on a prescription pad my question. So, I wrote, 'Zomax, does it cause anaphylaxis?' I did receive a reply by letter describing how safe it was, and that there were no major reactions that they knew of at that time."

But the April 16, 1981, issue of the highly prestigious *New England Journal of Medicine* had already published a report of an anaphylactic reaction associated with Zomax.[31] This was "the first signal of a problem to us," testified the FDA's then Director of Drug Experience, Dr. Judith Jones. McNeil undoubtedly also noted the article. The company had a "science information department that would actually do the monitoring of the literature," testified McNeil's medical director. The department put out an "alert list" of articles about McNeil products appearing in the scientific and medical literature. The alert list would go to a wide range of scientists and officials of the company including the medical director.[32]

After a third case of anaphylactic shock, in January 1982, Berneis again spoke to a McNeil representative: "He said that I should call the company . . . they would tell me how safe it was and that we were the only people having trouble. I talked with the physician there in charge and she told me about 200 cases of moderate to severe reactions that they had on record but mentioned nothing about deaths." After Berneis relayed this information to the McNeil representative, "He didn't say any more after that. He really looked as if he didn't have any idea that this was going on."

What McNeil Wrote to Doctors

By April 9, 1982, the scientists and executives at McNeil knew that Zomax was associated with anaphylaxis. McNeil sent out more than 200,000 copies of a "Dear Prescribing Physician" letter with a warning: "Anaphylactoid reactions to Zomax have been reported and appear more likely to occur in certain patients." Which ones? The letter "re-emphasize[d]" that people known to have had a bad previous reaction to aspirin should not take Zomax. The same applied, said the letter, to anyone who ever previously reacted to Zomax or any other NSAID, including Zomax's nearly identical twin, Tolectin. The letter added "Hypersensitivity upon reexposure . . . cannot be ruled out."[33]

That is a rather limp warning, compared with the following one: "Most patients have taken Zomax uneventfully. However, the development of hypersensitivity with intermittent reexposure cannot be ruled out. Consideration of this fact should be given when prescribing Zomax." This kind of warning, which might have discouraged some doctors from prescribing Zomax, was not used although it had existed in an earlier draft of the same letter.[34]

According to Patrick Seay's September 8, 1984, "Productivity" memo, the drafters of this earlier, stronger version of the letter took the language almost entirely from the suggestion of an FDA official, Dr. Harter. The internal memo states that Harter had previously requested an even stronger statement. It said that, for patients who had taken Zomax in the past and for whom the drug would be given again, the prescribing physician should "consider administering it under medical care."

"[S]o the caution about patients with a history of uneventful exposure to Zomax has been removed," said Congressman Weiss. McNeil denies this. The earlier draft is in the subcommittee transcript, however, and Dale and Lemanowicz, who claim to have worked on a total of seven earlier drafts "emphasizing uneventful intermittent exposure"[35] also say that the warning was "omitted."[36] In a January 1992 telephone interview with me, Lemanowicz explained, "The final version was not as recommended by the scientific and medical people who worked on earlier drafts of it. The pressure to change it was from marketing and salespeople." McNeil admits that the "drafting of the letter was a collective effort" but denies that the two scientists were given an assignment to draft it.

The wording was particularly weak in light of a March 31, 1982, memo sent to McNeil's president as well as others in the company, from Thomas W. Teal, director of McNeil Statistical Services. It warned that intermittent usage was the single predictable, identifiable risk factor linking Zomax with anaphylaxis in patients who had no history of allergy to aspirin, Zomax, or any other NSAID.[37] Teal analyzed the 178 known (to McNeil) allergic/anaphylactic reactions to Zomax from the adverse reaction reports. Of the 100 cases where the dose regimen was known, 68 reactions occurred during restarting the drug.[38] The FDA never saw the memo. What does McNeil say about this? The company "admit[s] that on March 31, 1982, Teal wrote a memorandum which was not formally submitted to the FDA as part of the NDA file." McNeil also denies that "the internal Zomax adverse reaction data on which it was based was not submitted to the FDA."[39] Lemanowicz said in an interview, "The format and the conclusions made in that memo were not submitted to the FDA."

In his September 8, 1984, memo Seay wrote, "[W]hen the various actions on Haldol [another McNeil drug], Tolectin and Zomax are taken together, it appears to me that a skilled lawyer may make it appear that we were less than energetic" in advising physicians about serious side effects.

He added that brand managers increasingly influenced both the wording and the alterations in the package insert, and so changes desired by medical and scientific people became harder to accomplish. "We resisted too much and waited too long in making changes in the package inserts for . . . Tolectin and Zomax."

In December 1982, Jack Yoffa, MD, a practicing gynecologist–obstetrician in Syracuse, New York, swallowed a Zomax just before driving to the hospital for minor surgery. "About half way into Syracuse on Route 690 . . . I started to have itching, pruritis. . . . I looked in the mirror, saw my face getting red. . . . I was about 4 minutes from my office and . . . I [figured I] will just try to get there and get treatment. Within 60 seconds of starting to itch I was totally unconscious, hit a guardrail . . . crossed a three-lane highway, miraculously missing several cars, struck down a light pole, and went into a ditch. . . . On arriving at the hospital in Syracuse, I had a blood pressure of 70 over zero, a pulse that was so fast that they really could not determine what it was."

How McNeil Acknowledged Deaths from Zomax

Yoffa began gathering information on other reactions to Zomax and by February 1983 had compiled a list of 10 cases just in the Syracuse area. He contacted McNeil, and the company doctor investigating the cases visited Syracuse to interview the patients.

According to Yoffa: "My wife and I joined him for dinner. . . . He seemed to be extremely concerned, had told me . . . about his discussions of Zomax with the people from not only McNeil but also Johnson & Johnson. . . . I did ask him twice, on two different occasions, if he had any reported deaths. His answer to me at that time was no. . . . He said that up until that time they only had scattered incident reports and nothing really very serious—no very serious reactions. . . ."

The McNeil doctor was among those to whom had been addressed an August 27, 1982, memo from McNeil's Zomax product director to James Dale concerning the importance of dealing with adverse reactions. The memo said that not only had adverse reaction reports been coming in to the company concerning Zomax, but also other sources had been bringing in unfavorable information, including existing and forthcoming articles in the scientific literature, remarks from medical doctors, salespeople, district managers, and Dale himself. Taken together they gave a "decidedly unfavorable picture." The memo added that the company did not have a thought-through program for coming to grips with the anaphylactoid "problem."[40] McNeil admits that this memo was sent to all the people listed on it as receiving copies.[41]

In fact, on March 12, 1982, nearly a year before, McNeil had gotten a report of a lethal anaphylactic reaction associated with Zomax.[42] "It was only several days after [the McNeil physician's visit] that they announced that they did have knowledge of deaths," testified Yoffa.

Within four weeks of starting his inquiries Yoffa had a list of six cases in the Syracuse area. "I presently have knowledge of between five to seven deaths, [other than those reported by the company]" testified Yoffa in April 1983.

In February 1983, the same McNeil physician visited Stephen Alexander, MD, a family doctor from the Syracuse, New York area, and a professor at the Upstate Medical Center in Syracuse, who had reported to the FDA three severe but not fatal Zomax reactions. Alexander testified, "He was very proper and very correct and seemed very concerned. . . . I . . . asked him if any deaths had occurred and he did not reply in the affirmative."[43]

That was the week of February 7, 1983, eleven months after the first reported death. On February 8, 1983, McNeil notified the FDA of a second death.[44]

In January 1983, Dr. Devra Davis, an epidemiologist, toxicologist, science policy director for the Environmental Law Institute, and professor at Johns Hopkins University, took a Zomax in the middle of the night because the pain from her broken foot had awakened her: "Within 23 minutes my pulse was 130. I awoke my husband . . . to get the *PDR*—the *Physician's Desk Reference,* the book that lists reactions to drugs. . . . There was . . . no indication that I could be having an anaphylactic shock reaction. . . . Apparently, I stood up then and fell down a flight of stairs. . . .[45] [After she came to in the emergency room] I asked if this had happened before, and one of the nurses said that she thought they had seen 12 cases that month."[46]

When McNeil "temporarily" withdrew Zomax from the market on March 4, 1983,[47] five deaths had been confirmed and additional death reports continued to trickle in. Based on the adverse reaction reports McNeil submitted to the FDA, by October 7, 1983, at least nine patients had died from an anaphylactic reaction associated with Zomax. The FDA couldn't conclusively determine whether Zomax was associated with another 28 deaths.[48] However, even these figures could have grossly understated the case, according to Dr. Davis: "One of the things we know in epidemiology, which is the study of disease in population, is that the actual recorded rate of deaths due to Zomax has to be far fewer than those that have occurred. . . . [I]t is very difficult to question people once they are dead about what they might have taken in the 20 minutes before they died."

Dr. Davis pointed out that autopsies may not clarify matters: "At autopsy, persons who die from allergic reactions do not necessarily appear different from those who succumb from cardiac arrest or respiratory failure. In fact, two weeks after my reaction, I removed hundreds of samples of Zomax from my grandfather's nursing home. Think of all those elderly people with heart or lung disease who die in their sleep and may have suffered an unrecorded anaphylactic reaction."[49]

Then there were those who survived but suffered very severe reactions. By the time McNeil withdrew Zomax, there had been almost 60 "truly life-threatening reactions," according to Dr. Robert Temple, the FDA's acting director of the Office of New Drug Evaluation. According to Dr. Temple, another 160 cases, while not evaluated as life threatening, were "bad enough" to require hospital care.

On February 6, 1983, President O'Brien received some bad news, according to Dale and Lemanowicz. They say that most of the MDs in McNeil's Medical Division told him they "no longer had confidence in the safety of Zomax" and "wouldn't use or prescribe it." McNeil admits that "several physicians at McNeil expressed concern over reports of allergic/anaphylactic reactions to Zomax and opined that revised labelling should be considered."[50]

McNeil kept Zomax on the market for another two months. Then, on March 3, 1983, a Syracuse TV broadcaster featured the Zomax story on his program when he interviewed Jack Yoffa, the doctor who had blanked out and driven across a three-lane highway. "Mr. Art Peterson from channel 9 . . . broke this story and . . . I think was responsible for literally, possibly in the next year or two years, saving thousands of lives. I am very indebted to Mr. Art Peterson." The next day, March 4, McNeil announced they were temporarily taking the drug off the market. That same day, McNeil representatives spoke on the phone with FDA official John Harter and another FDA scientist, Dotti Moore. According to Moore's memorandum of that conversation, the McNeil representatives told the FDA then that with reference to Zomax anaphylactoid reactions, "75% of the cases were with intermittent usage or restarts."[51]

By September 15, 1983, McNeil acknowledged to the FDA 2,161 allergic/anaphylactoid reactions associated with Zomax.[52]

As part of Johnson & Johnson's later efforts to remarket the drug, the company initiated a lobbying effort immediately after the April 1983 Weiss hearing. The hearings had been held to investigate FDA procedures in approving Zomax in the first place and then in monitoring the adverse reactions to it. William J. Ryan of the Johnson & Johnson Office of the General Counsel sent a November 4, 1983, memo to a number of top McNeil officials. It stated in part that the company had decided on a plan that involved making contacts with some members of the Weiss subcommittee. The memo mentioned two specific congressmen and said that the details of how the Zomax matter would be discussed with them was being completed.

The memo was attached to a six-page letter from Manfred Ohrenstein, New York State Senate Minority Leader, to Congressman Weiss. Ohrenstein wrote that in his view both McNeil and Johnson & Johnson had clearly displayed highly responsible actions toward the public concerning Zomax. As soon as the company found out about the "small number" of allergic and anaphylactic reactions out of 15 million Zomax users, the company reported the reactions at once to the FDA and withdrew Zomax from the market voluntarily.[53]

By contrast, medical practitioners in a McNeil-sponsored focus group did not feel the company was acting responsibly with regard to reintroducing Zomax. "This is dollars before lives," said one physician. Many of the physicians in the focus groups objected to the proposed use of a new label that "puts the onus on us."

A proposed label was presented to the focus groups that warned: "Patients starting therapy with Zomax 'should do so in a setting where adequate emergency treatment would be available . . .'" The Report noted that the focus group doctors were amused by this sentence since it seemed to require that Zomax could only be administered in a hospital emergency room because no place else is as fully equipped to help deal with an anaphylactic reaction. The Report also stated that the focus group doctors felt resentment toward the company for seeming to put all the accountability for careful use of Zomax on the prescribing physician. The doctors felt that lawyers write these proposed labels with the aim of shielding the company from, and as a result exposing the doctor to, charges of bad conduct.[54]

CONCURRENCY

Marketing Drugs and Medical Devices

Pressure on Sales Representatives. Zomax, as is commonly the case with new drugs, was still being researched after it reached the market. This is standard industry practice and, to some extent, is unavoidable.

One thing companies search for is a rare adverse reaction, or side effect. A former FDA Commissioner, Dr. Arthur Hull Hayes, Jr., explained: "An adverse reaction with a frequency of 1 in 1,000 patients is considered uncommon, but if it is medically serious, it clearly assumes public health importance if the drug will be used by millions of people. . . . The monitoring of adverse reactions is well-accepted today as an essential scientific endeavor that complements the pre-market evaluation of a drug."[55]

This "essential scientific endeavor," unfortunately, occurs in the worst possible context—during the frantic marketing and promoting of the drug. This is known as Phase IV of FDA approval.

Jody Perez was a McNeil sales representative who quit out of conscience: "With all pharmaceutical companies, any time a new drug comes out I would state there is a massive sampling. . . . [F]rom October 1981 until December 1981, we were going into the last quarter of the big launch for Zomax. . . . With my own personal territory I had a goal of leaving each personal physician . . . between a case to two cases of Zomax. Again, like I said, it is very competitive, especially with all the nonsteroidals. Each company and each representative is trying to get his drug across to the physicians, and sampling is just like advertising."[56]

A lot of pills are dumped on a lot of doctors. Each case contains 48 boxes and each box, 40 pills. So each physician contacted was handed a minimum of 2,000 Zomaxes, and some received nearly 4,000. Perez testified that he dealt with "about 190" doctors. Averaging this out to 3,000 Zomaxes per doctor, one sales representative gave away more than half a million doses of the drug.

The Tendency to Underreport Adverse Effects. In a high-pressure selling context, sales representatives, eager to push a new drug, rather than undermine it, may soft-pedal complaints of adverse reactions. The testimony shows that this is precisely what happened with Zomax.

The FDA, according to Commissioner Hayes, asks physicians to report bad drug reactions on a "drug experience report. The bureaucratic jargon is, a 1639. . . . It asks for a description of the reactions, the laboratory data, what drug is involved, were other drugs given, and how did the patient do. Is he or she alive?" The form is one page long; the FDA sends out millions of them each year to physicians. In this way such cases are recorded with the FDA. "[T]hen we take on the burden of following it up with more details, . . . [because] a reported reaction may not be valid."[57]

Few doctors, however, fill them out.

Dr. Sidney M. Wolfe, Director of the Public Citizen Health Research Group in Washington, DC, testified, "[S]omething like three-quarters or 80 percent of the average reactions which are reported to the FDA, come from the company, originally told about them by the doctor and hospitals. . . . Doctors are worried, hospitals are worried that even though the reporting [directly to the FDA] is entirely anonymous, that the reporting might somehow result in malpractice actions against them."[58]

Conflicts of Interest. As the testimony at the Weiss Zomax hearings shows, the drug company representatives have a conflict of interest inherent in their roles. They are simultaneously trying to push the drug and gather or pass along information that could discourage sales. A surgeon said the same thing with regard to medical devices, "It is patently absurd and inherently fraught with conflict of interest to expect the manufacturer of a medical device to be the sole source of reporting problems with that device in a fair and objective way when the reports are being given to the doctors who the manufacturers are simultaneously trying to convince to continue using the same product."[59]

The problem is clearly worldwide. The Professor of Cardiothoracic Surgery at Queen Mary Hospital, Hong Kong, wrote John Dingell in 1989 concerning the Bjork–Shiley convexo-concave heart valve, which had by this time caused many deaths: "I had discussed the issue verbally with the Shiley representatives on numerous occasions. Initially, they gave me an impression that I was the only surgeon seeing such problems. Later on, when they admitted that other surgeons also encountered this problem, they repeatedly stressed that it was very rare. They began to express some sympathy when this became a recognized problem worldwide."[60] A staff anesthesiologist at a Portland, Oregon, hospital, who was formerly a drug company sales representative, testified, "The primary goal and the primary measurement of the sales representative's success today as it was 20 years ago, is gross sales and percentage increase from last year."[61]

Tactics to Discourage Physicians' Reports. A medical practitioner reluctant to get involved helps further to tilt that conflict of interest. Testimony shows, however, that company representatives may use other tactics to discourage the reporting of adverse reactions.

For instance, a representative may try to intimidate doctors into thinking that they may be misinterpreting the adverse data. Dr. Yoffa testified: "In fact the McNeil representative who came in to me made me feel ridiculous by intimating that I should have asked these patients who had reactions if they had had an aspirin allergy, and that in all probability these patients did, and that is why they had a reaction to Zomax. In fact the incidence of true allergy to aspirin is almost negligible, and very few of the patients who have had reactions have ever been allergic to aspirin."

Doctors may also get the impression that the company feels they are exaggerating the problem. A Lubbock physician wrote the FDA, "We notified the manufacturer, McNeil, but the company did not feel four reactions within a 20 day period in one town was sufficient for alarm."[62] Dr. Berneis testified that after his second case, "This time we were made to feel as though we were the only people having this, and how could a small town area have two?"

Understaffing at the Food and Drug Administration

If the doctor only reports the reaction to the drug's manufacturer, even in writing, he may never know if the information gets to the FDA. In addition, the FDA cannot handle all the forms cascading in. Daniel Sigelman, a staffer for the Weiss subcommittee, wrote a memo about this problem:

"During my examination of the Zomax NDA, I found no sign that most of the volumes containing post-market adverse reaction reports had been reviewed by a medical officer [of the FDA]. It was my understanding that medical officers were generally expected to initial all reports which they reviewed. Hundreds of Zomax adverse reaction reports which I examined bore no such initials. . . .

"To prevent her office from being totally inundated with unreviewed volumes of adverse reaction reports, . . . [the FDA official] stated that she often returned them to the shelves in the Division's document control room. . . ."[63]

The government person in charge of oversight ends up using the manufacturer's data. The Sigelman memo explains how the fox ends up guarding the chicken coop:

"The backlog he [Dr. Harter, the group leader of the Zomax review] faced forced him, he said, to depend on periodic adverse reaction summaries submitted by sponsors for their drugs. . . . [An FDA official] had informed me that sponsors customarily submit quarterly reports of such summaries for the first year of marketing and yearly reports thereafter."

To Congressman Weiss's question, "Would it be fair to characterize your resources as being stretched to the breaking point?" Harter replied, "I don't think we broke, but we certainly were stretched." A hopelessly understaffed government bureaucracy relies on the data supplied by the manufacturers it "oversees." This is what Commissioner Hayes's "essential scientific endeavor" during the Phase IV research comes down to.

Slow Reporting by Manufacturers

McNeil is not the only company to hear from the FDA about sending in adverse reports on their drugs. Other drug companies have also been notified that they have not passed along adverse drug reaction reports (1639s) with the speed the FDA and most everyone else would like. For example, said Congressman Weiss, "Pfizer Pharmaceuticals had failed to report to FDA at least 26 serious adverse reactions associated with use of its arthritis drug, Feldene, outside the United States." On December 9, 1982, the FDA wrote Pfizer, "We feel such reports should have been available as a minimum to Pfizer's U.S. physicians and optimally to FDA reviewers as well during the deliberations about Feldene's safety and adverse reaction labelling."[64]

Some manufacturers of medical devices have also failed to supply the FDA with timely adverse reports, as shown in congressional testimony about the Bjork-Shiley Convexo-Concave Heart Valve:

CONGRESSMAN DINGELL: The FDA required as a condition for approval for the PMA that Shiley report to FDA within 10 days any adverse reactions, such as strut fracture. Did Shiley meet this requirement in early years?

MR. BENSON, ACTING FDA COMMISSIONER: I think they did not.

CONGRESSMAN DINGELL: They did not. Do you agree with that, Mr. Villaforth?

MR. VILLAFORTH, DIRECTOR, FDA'S CENTER FOR DEVICES AND RADIOLOGICAL HEALTH: Yes.[65]

At the same hearing, Roger Sachs, vice president of Shiley, Inc., testified that during the early years "[e]ssentially, the ground rules for many things, such as reporting, had not been established. I think that the gentlemen from the FDA were a little confused about something in terms of reporting to them."[66]

The FDA officials were testifying about a number of documented delays, which ranged according to a report issued by Dingell's subcommittee, "from 3 weeks to as long as 24 months." The report discussed the significance of these delays: "[S]ome of the delays coincided with critical junctures in the FDA's review and performance of the valve."[67]

A December 1990 internal, confidential FDA "Task Group Report on the Bjork Shiley Heart Valve and the Shiley Corporation" concluded: "During the c/c [convexo concave] valve's history, Shiley engaged in efforts to thwart FDA's intervention by untimely reports of fractures, unreported changes in quality control and manufacturing procedures, failure to correct known poor manufacturing procedures, and minimization of the overall problem through misleading and confusing communications to FDA and the medical community."[68]

Shiley has vigorously and repeatedly denied these allegations.

Problem with Reports of Animal Research

Concurrency in drug research is not limited to the reporting of adverse reactions. Too often, basic animal research testing goes on simultaneously with high-pressure marketing of drugs.

Three examples follow:

Dolobid (generic name *diflunisal,* a nonsteroidal anti-inflammatory drug). The manufacturer agreed to conduct a two-year mouse study concurrently with Phase IV marketing. The reason: the original mouse study was flawed. The 10 mice in the high-dosed group were killed after Week 85. "The killing of all the high dose animals at Week 85 might have masked the incidence of tumors in this group," said the FDA's statistician. This was particularly important because of what happened in the medium-dosed group. Of the 15 tumors in the mice in the mid-dosed group, 6 only showed up between Weeks 86 and 97. "The incidence of lung adenomas in the medium dose male mice is significantly higher than that in the controls," the statistician added.[69]

Oraflex (generic name *benoxaprofen,* another NSAID). The manufacturer had arranged for two animal studies that would be concurrent with Phase IV marketing. One was a two-year mouse study that was actually underway, and the other was a one-year rat study for which the protocol had been submitted. Why all the animal studies? At the time the FDA had approved the drug, the only finished, prior study on carcinogenicity was a two-year study on rats. This would be fine, except so many treated rats died that the results couldn't be given meaningful statistical analysis. The FDA's pharmacologist gave a chilling conclusion, "The two-year rat study does not support the safety of this drug for its chronic use in humans."[70]

Feldene (generic name *piroxicam,* another NSAID). The FDA told the manufacturer "that a commitment to do a [postapproval] 24 month animal study would be considered adequate for approvability and would not slow the approval process." Why was such a study needed? The FDA's pharmacologist wasn't sure whether earlier 18-month rat and mice studies lasted long enough to show that the drug was safe for long-term use.[71]

These examples suggest that the FDA "has approved drugs without the data needed to 'rule out a major risk.'"[72]

Inadequate Testing of Medical Devices

Fractures in the Bjork–Shiley Convexo-Concave Heart Valve. John Dingell's Oversight and Investigations Subcommittee Report made the same point about the Bjork–Shiley Convexo-Concave heart valve. According to the report, Shiley put the valve through at least five major manufacturing changes between 1979 and 1984. In June 1979, the manufacturer increased the angle of a key strut to overcome a tendency toward fractures, but the new angle was unsuccessful. In May 1980, the company started a new load test on the strut to try to sort out good struts from bad ones, and this sorting process also proved unsuccessful. In February 1982, Shiley stopped using that test, initiated a new final inspection test, and also changed the manufacturing procedure for lifting the strut hook into place. Still the fractures continued.

By March 29, 1982, even Dr. Viking O. Bjork, who had worked with Shiley to design the valve, was losing patience. He sent a memo to top Shiley managers: "Have you not yet realized that strut fracture is one question brought up wherever I appear. . . . You're [*sic*] circling around with other solutions is probably a waste of time. At this stage welding will not be acceptable . . ." He went on to state that the manufacturing process was "unacceptable" and that Shiley had not given him "trustworthy data" about what they planned to do about it.[73] The fractures continued. A year later, in July 1983, Shiley started a new weld inspection, but the struts continued to fracture. In March 1984, the company tightened the tolerance for the strut location.[74]

Republican Congressman Michael Oxley asked James Benson, the Acting Commissioner of the FDA, who testified in February 1990, "The officials from Shiley apparently have said that they feel they have the strut problem solved. Do you agree with that?"

Benson replied, "I think that—no, I do not agree with that, to give you a simple, direct answer. I think—as I understand it, there have been no new reported strut fractures since April or sometime in 1984. That's not long enough to make that determination. I hope the latest modifications that were made solved it, but we do not know that yet."[75]

Some of the fractures may have been caused by a "phantom welder." A number of valves were rewelded when inspectors detected problems. However, in some instances, the rewelding was not done, and instead the worker simply polished out the crack. The FDA's December 1990 confidential Task Force Report described information on this obtained from interviewing plaintiffs' attorneys in a Houston personal damage suit, "To conceal this activity which was contrary to established procedures, the records reflecting the history of the device were falsified to indicate that rewelding had actually occurred. The falsified records contain the employee number of an individual who was not in the employ of Shiley when the records were generated. This so-called phantom welder appears on records as having rewelded cracks that had been detected by the quality control inspectors. As many as 1,700 valves are reportedly involved." Pfizer, the parent company, has been investigating but as of November 1991 had not determined who the actual person was who did the rewelding. "We do not yet know all the facts surrounding the . . . records, which were made a decade ago," said the company in an official statement.[76]

Undocumented Advantages of the Bjork–Shiley Convexo-Concave Heart Valve. The FDA had concluded in 1979 that the Convexo-Concave valve threw off fewer tiny blood clots than did Shiley's previous valve, giving the new valve a clear clinical advantage,[77] but the evidence for this supposed advantage was inconclusive. The original package insert shipped with the valve states, "Professor Bjork has not experienced early thrombosis with the Convexo-Concave valve but suggests that longer follow-up is necessary to determine if the incidence of valve thrombosis and embolism has been significantly reduced."[78] Even the valve's package insert acknowledges the results on the new valve were not in. Yet it was being manufactured and sold.

By June 13, 1984, the FDA no longer believed the new valve was any better than the older one. The FDA wrote Shiley, "After nearly 8 years of clinical use, the convexo-concave 60-degree valve does not appear to be statistically better overall than the spherical valve when experience with thromboembolism, thrombosis, and strut fracture in the reports are analyzed."[79] As a result, the agency threatened to reverse the valve's premarket approval unless Shiley could produce evidence of the device's superiority. Shiley brought in Dr. Bain, a Scottish cardiovascular surgeon to a July 10, 1984, meeting with the FDA. Dr. Bain was sort of a one-man laboratory, having implanted 668 of the earlier Shiley-Bjork valves and 442 of the convexo-concave ones.[80]

Using data Bain had accumulated, Shiley won over the then FDA Deputy Director of the Center for Devices and Radiological Health, James S. Benson. However, as the FDA stated in a task force report, "[T]he firm itself knew its data did not support such a conclusion [that the convexo concave valve was superior]." An internal, July 6, 1984, memo from Pfizer to Shiley, which Benson had not seen at the time, stated that the data submitted to the FDA "is not really an adequate proof" to distinguish the two valves. The memo goes on to question the "possibility" of obtaining adequate data."[81]

Benson did not know about this memo, and wrote in a July 16, 1984, letter that the advantages of the valve in reducing tiny blood clots normally thrown off by artificial heart valves outweighed the strut fracture problems.[82] When asked later if he regretted signing the letter, Benson testified, "That conclusion, I think, was wrong; therefore, I regret it."

Benson's letter required Shiley to send the actuarial data

gathered by Dr. Bain. Shiley had used one version of Bain's data at the July 10, 1984, meeting with FDA officials. However, the data could be analyzed by other statistical methods, and an FDA statistician, Dr. Harry Bushar did so. He wrote in an August 1985 memo: "Based on Bain's data alone as of August 30, 1984, the Bjork–Shiley Convexo-Concave 60-degree valve no longer appears to be statistically superior to the Bjork–Shiley Standard valve when valve thromboses and systemic emboli are compared. At least one more set of Convexo-Concave and Standard data should be analyzed by the actuarial methods since Bain's data appears to have changed drastically when subjected to the critical review required for the actuarial method."[83]

Bushar kept analyzing Bain's data for two years, concluding, according to the subcommittee report, "that there was no statistically significant difference between the two valves." Acting Commissioner Benson described this statistical analysis as "ground-breaking research." But if either the FDA or Shiley or both were doing ground-breaking research, Shiley continued to sell the valves worldwide during this analysis. By 1983, Shiley had half the world market share for mechanical heart valves, and half of those sales were convexo concave valves.[84]

Dingell's subcommittee report concluded: "Shiley, Inc.'s decision to continue to market the valve despite its unsuccessful attempts to identify or correct the cause of the strut fractures illustrates a common practice of conducting clinical trials while continuing full scale marketing. This 'earn as you learn' philosophy has prematurely cut short the lives of hundreds of implantees. In some cases, the Subcommittee has learned that some valves fractured only months after being implanted."

These examples show conclusively that the dreadful concept of concurrency is common in corporate medical research.

SECRECY

McNeil's Handling of Zomax Warnings

Although McNeil voluntarily made a "temporary withdrawal" of Zomax from the market, with the intention of remarketing it, neither the company nor the FDA ever issued a formal recall.[85] Perhaps as a result, not all doctors seem to have gotten the word, and a lot of Zomax continues to sit in people's medicine cabinets.

A recent article in the *Lancet* reports on three post "temporary withdrawal" deaths associated with Zomax—a 38-year-old man in June 1984, a 37-year-old woman in December 1984, and a 28-year-old woman in April 1985.[86] Letters were written to McNeil in two of these instances advising them of the Zomax-related deaths, but these documented deaths did not elicit from the company any warnings to the public about the dangers from this drug.

A March 17, 1983, memo from a public relations officer at Johnson & Johnson suggests that there is no need for further warnings because the product has been taken off the market.[87]

With so many cases of anaphylactic shock, McNeil was the target of 600 lawsuits after withdrawing Zomax.[88] McNeil settled virtually all of them before they ever got to court, and as part of many of the settlements, persuaded the judge to issue a protective order. According to attorney Richard Campbell, who, as reported on his own biographical sketch, "substantially limits his practice to the defense of major product liability cases": [Protective orders] typically order the plaintiff and his attorneys, experts, and consultants to forgo use of the confidential information and documents disclosed to them during the discovery phase of a case for any purpose other than advancement of the specific litigation and to return the documents to the defendant at the end of the case.

Since McNeil settled most Zomax cases out of court, few relevant documents became a matter of public record. Dale and Lemanowicz had not settled by early 1991, and the documents they and their attorney obtained in discovery became publicly available—in a federal courthouse. Their attorney Allan Kanner, in fact, successfully fought off several attempts by the company's lawyers to impose a protective order early in the proceedings.

Withdrawal of the Bjork–Shiley Convexo-Concave Heart Valve

As a consequence of protective orders, the Bjork–Shiley convexo-concave heart valve may have caused a sizable number of fatalities. An FDA Compliance Officer testified, "Shiley is producing a heart valve that demonstrates a nearly unique failure phenomenon—strut fractures. . . . Well over 100 fractures have thus far been reported, with a high percentage resulting in fatalities. . . ."[89] Added the Acting Commissioner of the FDA, James S. Benson, "[T]he fracture rate is real. It is above most other valves."

By the time the company withdrew the valve from the market, on November 24, 1986, more than 85,000 of these valves, with their potential for failure had been implanted, worldwide, into people's chests.[90] The FDA's confidential Task Force Report explained, "The firm maintained that it [withdrew the valve] not because of concerns about the performance of the c/c valves, but because market conditions had made further production essentially uneconomical."

People walking around with the valves implanted in them did not always know there was a potential problem; in lawsuits, there had been "protective orders on all cases," according to Bruce Finzen, a well-known plaintiffs' attorney. The husband of one of those implantees testified: "My name is Fred Barbee. I am from Monong, Wisconsin. My wife Carol, who was fifty years old at the time, had a Bjork Shiley 60-degree heart valve implanted on May 26, 1982. After 6 years, that valve failed due to a strut fracture. Carol died approximately 48 hours after she first showed symptoms of what I now know to have been the valve failure, after there were great efforts to keep her alive. The symptoms of the valve fracture are much like a heart attack, and because we had never been advised about any sort of valve failure whatsoever, I made an incorrect decision to take her to the closest, but limited, facility that could treat a heart attack, but not a broken valve. Because of the lack of that same information an incorrect diagnosis by the emergency room physician . . . was also made. That incorrect diagnosis and subsequent treatment caused Carol to slip into clinical death before we could get her to the appropriate facility. . . . had we been given half a chance, we could have prevented her death. . . . We should not be kept ignorant about the potential failure of a mechanical medical device, whether it be a heart valve or anything so vital to life itself."[91]

Barbee added in other testimony: "[T]he manufacturer of the valve, Shiley Incorporated, and its parent company, Pfizer, Inc., made a conscious effort not to directly provide that information to us and made the decision not to provide that information to my wife's regular treating doctors, deciding instead to only provide some information about valve malfunction to heart surgeons. However, my wife, like most open-heart surgery patients, had not seen her heart surgeon since the valve was implanted in 1982. The only other source of information available to us, the media and the press, was absent, I believe, because of secrecy orders and confidentiality agreements."

The surgeon who implanted the valve in Carol Barbee's heart did in fact get the information on the valve's strut fracture in a letter from the company, according to Fred Barbee's testimony, and he did communicate it to the Barbee family—after Carol was rushed from the first hospital emergency room in Wisconsin to a second one in Duluth, Minnesota, where he had moved his practice. Shiley had sent letters about strut fractures to heart surgeons as early as February 5, 1980, two years before Carol Barbee's heart valve operation, although we do not know if her surgeon received one.

Shiley's limited distribution of its warning letters to doctors on the valve caused international problems, as shown by a portion of an October 22, 1985, letter from J. Van Linden, the Permanent Undersecretary of State for Public Health of the Netherlands: "I regard the fact that the information about the problems with the heart valves was supplied exclusively to cardiopulmonary surgeons as a serious shortcoming, as a consequence of which other physicians have been unable to get this important information at all or to get it in time."[92]

By December 1990 Pfizer, the parent company, launched a program to contact the estimated 55,000 living implantees to warn them of the danger. One problem was finding them. The company only had the relevant registration cards for half the patients who had received the implants, and much of that information might have been out of date or inaccurate. Pfizer made arrangements with Medic Alert, a California-based foundation concerned with medical information to track down these and the other patients to send them Pfizer's warning letter and keep them posted on information concerning the valve. According to the director of the FDA's Office of Training and Assistance, however, the letter Pfizer sent to doctors "has the appearance of slick promotional material for a new product rather than a notice about an important public health issue. I'm very concerned that many office managers may discard the package as simply junk mail."[93]

Settlements with protective orders are obviously perfectly legal. We now know, however, that people died, arguably because of inadequate public information about what can happen with Zomax or the Bjork–Shiley convexo-concave heart valve. The dangers of a drug or medical device—the problems with its science—can become private scientific property.

Not everyone thinks that this secrecy, which may be good for the corporate bottom line, is good for science in general. Dr. Devra Davis, the epidemiologist who nearly died after taking Zomax to relieve pain from her broken foot, testified: "At its heart, science is an inherently democratic institution, fueled by shared, freely exchanged information. Democracy rests on the informed consent of the governed. Hiding information about matters of health and safety imperils any democracy. As my own experience testifies, such secrecy endangers lives, perverts science and ultimately undermines democracy itself."

There is evidence of another type of corporate secrecy in the Bjork–Shiley heart valve case. In 1980, Dr. Viking Bjork of the Karolinska Hospital, Stockholm, Sweden, was planning to publish in the medical literature results he had gathered concerning the strut failures of the valve he had developed. On December 17, 1980, the company, Shiley, sent Bjork a telex: "We would prefer that you did not publish the data relative to strut fractures. We expect a few more and until the problem has been corrected, we do not feel comfortable." The telex added that company officials wanted to talk over the strategy for publishing the information during Bjork's forthcoming visit.[94]

In a February 1990 hearing of John Dingell's Oversight and Investigations Subcommittee, Oregon Congressman Ron Wyden said, "I read something like this, and this doesn't sound to me like it's in the public interest." He asked the president of Shiley during the period in question, "At any point, then, between 1980 and 1982, did you tell Dr. Bjork that he could publish this data?" The president couldn't recall, but his accompanying lawyer said they would submit the answer for the record. Here's what they submitted: "Shiley was unable to locate any document in which Shiley communicated with Dr. Bjork regarding his publication if [*sic,* of] Shiley data in the period through 1982. In fact, however, in public lectures and written papers, Dr. Bjork frequently discussed strut fracture. For example, on September 10, 1982, at the Joint International Cardiovascular and Thoracic Surgical Conference in Stockholm, Sweden, Dr. Bjork included in two lectures a discussion of strut fractures, with photographs of fractured valves.

"It should also be noted that, as information was received and analyzed regarding strut fractures, Shiley sought to keep FDA and the medical community informed."[95]

At the hearing itself, former Shiley president Bruce Fettel testified, "[W]e had heard about a couple of other fractures, we were trying to chase it down and see if, in fact, they were real, and make sure that when Bjork published, that he would include all of the data . . . and not just represent it as being a small problem.

"So although it reads like we were trying to keep him from publishing, in fact, we were trying to keep—get him to publish the entire story. So it is completely misleading."

FOREIGN SECRETS

Approval of Versed in the United States

The major pharmaceutical companies are multinationals, and often market a drug in Europe before it is introduced into the American market. The drug company, however, may keep secret from the FDA critical medical knowledge on a specific drug gained from its experience in Europe. A case in point was discussed by Congressman Weiss in a 1988 investigation of the anesthetic and sedative Versed (midazolam hydrochloride), manufactured and marketed by Swiss-based Hoffman–La Roche. Versed is used for conscious sedation in surgical procedures, sometimes in doctors' and dentists' offices rather than in hospitals, and was marketed in the United States on March 19, 1986.

At his May 5, 1988, hearing, Congressman Weiss said, "Subcommittee staff review of records obtained from Hoffman–La Roche indicates that a total of at least six deaths involving respiratory depression associated with foreign marketing of Versed were known to Hoffman–La Roche by approval time, but were not reported to FDA until after approval."[96]

Five deaths had occurred in West Germany, one in Britain. The transcript of the hearings documented the details of each case. Between July 1984 and December 1985 the Hoffman–La Roche affiliates received the report of each death. According to the subcommittee investigation, however, that is not when the company reported all but one of them to the FDA. (One death had been "only briefly summarized in July 26, 1985, submission to the" FDA, according to the subcommittee report.) As for the more detailed notifications, Roche reported one death to the FDA on May 1, 1986, four on June 3, 1986, and one on August 27, 1986. The June 3rd date is five months after FDA approval of the drug but only four days after a German medical publication *Deutsches Arzteblatt* published a warning to doctors that, in translation, warned, "*Take care when giving Midazolam* [Versed's generic name]! . . . in one case breathing stopped and in some cases cardiac arrest and ventricular fibration occurred. . . ."[97]

Following standard procedures, the FDA sent Roche a letter on November 8, 1985, telling the company that the agency was about to approve the drug. Then on December 20, 1985, the FDA sent the company the actual approval letter. After the company gets the first letter, but before the FDA sends out the second one, FDA regulations require that the company inform the FDA about "new safety information learned about the drug that may reasonably affect the statement of contraindications, warnings, precautions, and adverse reactions in the draft labelling." Did deaths meet the requirements of the statutes? Gerald A. Faich, MD, director of the FDA's Office of Epidemiology and Biostatistics affirmed, "[Y]es . . . deaths from this drug were reportable."[98]

Evidence before the subcommittee included one doctor's testimony of a 70-year-old woman's Versed-related death:

"This was done in the hospital. The patient was very carefully monitored. We had two nurses and there were actually three physicians in the room. One nurse, whose primary responsibility was to be at the patient's head totally communicating with the patient, determining patient sedation, determining patient comfort.

"The urologist and myself were at the patient's side and a nurse was with me utilizing equipment that may be necessary to help make the diagnosis. . . .

"The patient was initially sedated with about two milligrams of Versed, given intravenously. After a short period of time, the patient was still very alert and we gave her another milligram and a milligram, so she had a total of four milligrams and we began the procedure. This procedure is uncomfortable.

"She had a process called retroperitoneal fibrosis. . . . In the process of inserting this flexible instrument through a very fixed rigid piece of bowel, the patient experienced more discomfort.

Actually the patient requested more comfort and it was obvious that she needed it. One more milligram of Versed was administered. . . . On the way up toward the kidney, the patient experienced a greater amount of pain and, again, begged for more comfort and we responded to it. We gave her one more milligram of Versed. . . . The nurse noticed that all of a sudden the patient quit breathing. She said, Dr. Walta, something is wrong. At that point in time, within less than a second, the instrument was removed and it was noticed that the patient was having what looked like a seizure.

"With that [the] cardiac arrest [team] was immediately called. . . . The patient was immediately intubated [The anesthesiologist places a tube into her breathing channel and breathes for her] but for reasons that none of us could understand although everything was done correctly during the attempt to resuscitate her, the patient continued to turn bluer and bluer. . . . We had somebody beating on her heart to put blood through her system. . . . Medications were given to try to improve the oxygenation, to improve the heart rate, and all of it was to no avail. Before our eyes, we watched this woman die."[99]

Patients were dying because their doctors were innocently overdosing them. Versed, which was about three to four times more potent than Valium (also manufactured by Hoffman–La Roche), was marketed with packaging and labeling that could lead physicians to believe it was about the same strength as Valium. A practicing physician explained to the Weiss subcommittee: "One point is the way the dosage vial comes out. In Valium it came out 10 milligrams in one vial. You draw up the 10 milligrams in a syringe, and that's usually the standard dose. When Versed came out, it came in a 10-milligram per vial or syringe. You drew it up in the syringe, 10 milligrams."[100]

A reason for packaging Versed like the much less potent Valium was revealed in a tape recording of a company official addressing the company's board of outside anesthesiologists on October 18, 1986, that was played at the Weiss subcommittee hearing. The speaker was identified only as Mary from Roche,[101] who said that one of the company's major considerations was to get Versed into the hands of anesthesiologists "before generic diazepam [i.e., Valium] became a stronghold . . ."

Mary from Roche described the effect of the company's marketing effort, including, of course, the packaging. To applause from the board of outside anesthesiologists she noted that, with the assistance of the assembled doctors, sales had exceeded their target, reaching $18.7 million and with 67.2% formulary acceptance, the "highest" formulary acceptance the company had ever accomplished.[102]

The original FDA-approved recommended dosage for Versed, which was included in the packaging, was as follows: "Generally 0.1 to 0.15 mg/kg [milligrams of Versed per kilogram of body weight] is adequate, but up to 0.2 mg/kg may be given, particularly when concomitant narcotics are omitted."[103]

As Dr. Sidney Wolf explained: "A typical weight of a person—some weigh more and some less—is 70 kilograms, which would mean that the original dose would be 7 milligrams . . . to 10.5 milligrams. . . . It is the combination of the range recommended at 0.1 to 0.15 plus you could go as high as 0.2 that led everyone to believe that this is really in the same ballpark as Valium."[104]

Response Curve of Versed

Versed not only is much more potent than Valium but also has less tolerance for error because of its "dose response curve." Dr. Robert M. Julien, formerly Associate Professor, Department of Anesthesiology and Pharmacology, Oregon Health Sciences University, explained: "A dose response curve is a graph . . . to illustrate . . . that when you increase a dose of a drug, you increase its effect. With Valium, as you increase a dose over a fairly large range, you only get slight increases in sedation. You can increase a dose of Valium perhaps fivefold or sixfold in a patient and still not put them at great risk. . . . with Versed . . . very modest increases in dose . . . a two-fold increase . . . will take a patient from very light sedation to . . . a state of general anesthesia. . . . they are apneic, they are unresponsive."[105]

According to the Weiss subcommittee, the company conducted clinical studies before FDA approval, indicating the severity of the dose response curve for Versed, but did not show the results to the FDA until "several months after approval."[106] Roche submitted a protocol to the FDA for a study by Paul F. White of the Stanford University Medical Center on April 13, 1982. The researchers completed the clinical portions of the study "some time in the first half of 1984," said Congressman Weiss. "That probably is correct, yes, sir," testified the relevant FDA official.[107]

The researchers completed a statistical report on the study on November 28, 1984, that stated: "With respect to degree of sedation, midazolam patients had a significantly higher incidence of difficult or no response to commands after the initial dose of test drug ($p < .01$) and ketamine ($p < .05$) than diazepam patients. Each successively higher dose of midazolam resulted in a significant increase in the incidence of no response to commands after the initial dose of test drug." The responses for both midazolam (Versed) and diazepam (Valium) were measured before the administration of the third drug mentioned, ketamine, which the researchers used in the study to help make sure the patients kept breathing.

"The new and most significant finding" of White's study, according to Roche's own Versed clinical investigator, Dr. J. G. Reves, is that Versed "has a steeper dose-response curve than diazepam. This means that there may be less room for dosing error with midazolam."[108] Reves was actually discussing a May 1988 paper by Paul White and colleagues. The subcommittee report stated, however, "the paper published by White et al. in May 1988 was *based on the earlier data analyzed in this Roche report of 3 1/2 years earlier*" (italics in original).[109] The drug company sent the November 28, 1984, statistical report on the White study to the FDA on May 4, 1988, according to a written response to the Weiss Subcommittee from the FDA. "All of the important findings of the 1984 statistical report appear again in Roche's final clinical report to this study," said Weiss. This was completed September 4, 1985, but was not sent to the FDA until September 26, 1986, about nine months after FDA approval.

"Was the White study included in the Summary Basis of Approval for the drug?" asked the congressman.

"I don't think so," replied the FDA official in charge of these matters.[110]

By October 18, 1986, the company knew of more than a dozen deaths associated with Versed's use, but this didn't stop the laughs at the meeting that day of company officials and Roche's outside anesthesiologists on its advisory board for Versed. Here are more selected excerpts from a transcript of that meeting.[111] An anesthesiologist asked, "What was your sales goal, Mary?" Mary from Roche replied that it was $15.2 million for three quarters of the year, and that the company thought it might actually reach $26 million. As recorded in the transcript, Mary got some laughs from the assembled anesthesiologists when she said that this money had been made even though Versed was "more potent than we expected." She said that had the company known how successful the product would be, their sales goal would have been higher, perhaps approaching $40 million.

An anesthesiologist then said, "Mary, I'll tell you now. We all knew it wasn't 2 to 1 [i.e., in potency relative to Valium], but we didn't tell you."

Mary from Roche got more laughs from the assembled anesthesiologists when she replied to this physician that she wouldn't forgive him for keeping this information from her. She then acknowledged that proper dosage was their "major problem." She acknowledged that there have been instances of "oversedation, agitation," and that Versed was "more potent than was reported."

Versed Dosages in Europe

Years before FDA approval here, the potency of Versed appears to have been understood by Hoffman–La Roche in Europe. For example, in England, "since December 1982," according to Weiss, the recommended dosage printed on the drug's label was .07 mg/kg, considerably less than the 0.1 to 1.5 mg/kg recommended in the United States. The British label, with a 1982 date, is reproduced in the subcommittee transcript.[112] The British regulatory authorities approved their lower level dosage based on studies published in Britain in 1982 and 1983. Roche knew about these studies because they included at least two of them in their submissions to the Investigational New Drug Application (IND) file of the FDA in 1983 and 1984. This FDA file tracks animal studies to determine if the drug is safe enough for human tests; then if the FDA agrees to studies on humans, it is also the file that first receives the human clinical studies. Once the manufacturer has completed these studies, all the animal safety data, human clinical studies, pharmacological data, chemistry studies, and manufacturing information are sent to the FDA in support of the New Drug Application (NDA). "It takes two to three years for the FDA to review this mountain of information," said Ed Lemanowicz, who used to work on NDAs at McNeil Laboratories and now does so for a Japanese company entering the U.S. market. When Roche submitted the 1982 and 1984 studies to the IND file for Versed, the company had already applied for an NDA. This raised a question from Congressman Weiss to the relevant FDA official. "Once an NDA is filed," asked the Congressman, "isn't most agency attention focused on the NDA rather than the IND file?" Replied the FDA official, "Yes, sir."[113]

Even the Head of Medical Affairs for Roche Products, Ltd., wrote in the May 19, 1984, issue of the *British Dental Journal,* that "with the double potency [of Versed compared to Valium], it need not be difficult to over-sedate the patient. . . . I can confirm that we are looking to see whether further benefits could ensue from a more dilute solution being made available."[114] In fact, the British affiliate did add to its available products in the United Kingdom ampules containing only 2/5 the dosage of the original ampules. They first marketed these new ampules on February 4, 1985, ten months before FDA approval in the United States. They didn't notify the FDA about these weaker ampules until September 30, 1986.[115] In fact, in October 1984, the British affiliate even defended itself in a case involving a death of a 76-year-old patient by saying, "This patient was given a dose of HYPNOVEL [Versed's name in Britain] 4 times higher than the recommended dose for a person of his age."[116] The dose he was given, 10 milligrams, was approximately within the range of the initial recommended dose in the United States. The original U.S. label stated, "Patients 60 years old or older may require doses lower by about 30% than younger patients." As noted earlier, however, the label also stated that dosage could go as high as 0.2 mg/km. For a patient weighing 70 kilograms, even with the 30% dosage reduction, this could imply a dosage of nearly 10 mg.

The drug's West German label, before FDA approval in the United States, also warned against the U.S. dosage later recommended: "[d]oses higher than 0.1 milligram per kilogram of body weight may produce oversedation. . ." The pre-U.S. approval Swedish label also recommended a lower dosage than that suggested by the U.S. packaging in 10-mg vials and the U.S. label: "Normal dose 0.15–2.0 mg body-weight."[117] Did the company inform the FDA about these discrepancies in the labels before receiving U.S. approval? "To the best of my knowledge, no," said the relevant FDA official.

CHAPTER 6

CORPORATE CASH: CONFLICTS OF INTEREST IN RESEARCH FUNDING

INTRODUCTION—THE BLUESTONE INVESTIGATION

A congressional report notes, "The most widely publicized cases of scientific misconduct in recent years have tended to involve physicians conducting biomedical research."[1]

One of the chief causes of scientific misconduct is conflict of interest. Princeton Professor Patricia Woolf elaborated on the issue at a congressional hearing, "Scholarly research must be disinterested . . . neither the design of experiments nor reporting of the results of inquiries should be affected by the desires of the sponsors . . . it is particularly important that those who conduct research should not have a personal or financial stake in a particular outcome."[2]

Getting promoted or tenured at a major research university requires the production of cited publications. Generally, publications are cited when they produce original, positive results. So the reward system in academia sets up an inherent conflict of interest: ". . . pressure to publish research, in order to obtain a better job or promotion, could influence a scientist to misrepresent or fabricate data in order to publish more articles," stated the congressional report mentioned above.

This chapter focuses on research involving the clinical testing of new drugs and established drugs that are being evaluated for new uses. Such testing is often farmed out to university researchers to "assure" the objectivity of the research. Some of these studies are paid for entirely by corporations; others are paid for by government agencies. Sometimes the funding comes from both sources.

Conflicts of interest can be particularly troublesome in clinical testing. The congressional report quoted earlier noted that biomedicine and biotechnology can offer financial rewards in the form of consulting fees and honoraria for misrepresenting or fabricating data.

In the case that is the main focus of this chapter, Dr. Charles Bluestone, the principal investigator on a grant-funded study, was accused of misrepresenting data by Dr. Erdem Cantekin, a coinvestigator. Bluestone then blocked Cantekin from publishing his version of the research results. As Congressman Ted Weiss elaborated at a congressional hearing he held on the case, "The whole issue as I look at the record, is over who has proprietary rights over a study and data funded by the Federal government, with the university playing a complicit role in keeping information from the public that differs from the principal investigator's. . . ."

The focus on publication makes this case particularly worth looking at closely. Although the facts in the case are unusual, the documents they have produced strip away the glossy covers on major scientific publications to reveal how the scientific publication process works.

OTHER CASES OF CONFLICTS OF INTEREST

Retin-A

The case involving Dr. Charles Bluestone is by no means the only instance in which making money appears to conflict with doing good science. A report put out by Weiss's Subcommittee on Human Resources and Intergovernmental Relations details a number of cases of conflict of interest. In one case, the *Journal of the American Medical Association* (*JAMA*) published an article on a vitamin A derivative, with the brand name "Retin-A," that could *reverse* wrinkles in skin. People could physically look younger, according to the article, by smearing on an ointment containing Retin-A. In the same issue, the journal published an editorial titled, "At last! A Medical Treatment for Skin Aging." Johnson & Johnson, the manufacturer of Retin-A, held a press conference about the breakthrough. Retin-A was already available by prescription for acne, and with the publicity surrounding its anti-aging qualities, it quickly vanished from behind the pharmacists' counters. Sales jumped from a little more than 200,000 tubes a month to well over a million tubes.

The *JAMA* article stated that the study was performed "double-blind." According to the Weiss subcommittee report, however, "the senior researcher on the *JAMA* article admitted that the study was not double-blind, as required to avoid bias, because the researchers were able to determine who was receiving Retin-A by the inflammation it caused."[3] About a year after publication of the *JAMA* article, *Money* magazine pointed out that Johnson & Johnson had paid honoraria to some of the researchers on Retin-A before *JAMA* published the article and before the press conference took place. The Weiss subcommittee obtained Johnson & Johnson's financial documents to check the accuracy of *Money* magazine's claims, and found that "several scientists who served as spokespersons on behalf of Retin-A had major financial relationships with Johnson & Johnson. . . . Dr. John Voorhees, an author of the *JAMA* article, received $253,120 in research grants predating the Retin-A press conference, and an additional $687,750 during the following year. He also received payments as a Johnson & Johnson consultant in 1988 and 1989, and 13 honoraria payments in 1988, including the Retin-A press conference."[4]

A different scientist wrote the *JAMA* editorial on Retin-A, but she also received $3,600 in honoraria from Johnson & Johnson before the editorial came out, and $9,000 the following year. This is a small amount of money by the standards of medical research, but according to the report, her department at Boston University also received "$393,380 in Johnson & Johnson research grants during the 2 years prior to the editorial, and four grants totalling $185,406 during 1988–1989."[5]

Does Retin-A, in fact, work? Not necessarily. The National Institutes of Health (NIH) later concluded that "the safety and effectiveness of Retin-A has not been established." The Head of the

Dermatology Department at the University of Illinois at Chicago, Lawrence M. Solomon, has testified: "I believe there is a direct tie between the irritation the drug produces and the swelling which temporarily causes wrinkles to 'disappear.' If the harmful effect is that which makes you look better, one might just as well slap oneself silly and achieve the same result."[6] In fact, Retin-A may cause serious harm. Commissioner David A. Kessler, MD, of the Food and Drug Administration (FDA) has testified, "Retin-A is a photosensitizer which, if used chronically, could potentially increase the risk of skin cancer."[7]

The TIMI Trials

A second case illustrates a different kind of conflict of interest: researchers' ownership of stock in companies that manufacture the drugs being tested. The testing was arranged to see which of two "thrombolytic agents" could better dissolve blood clots after heart attacks. The research was described as "Thrombolysis in Myocardial Infarction" and, for obvious reasons, was always referred to as the "TIMI trials." One of the agents, streptokinase (SK), manufactured by Hoechst and others, had been approved by the FDA for intracoronary administration in 1982. The other, *t*-PA, manufactured by Genentech, was a new drug that had not yet been approved. The clinical trials, which began in 1984, involved 13 different clinical centers and five core laboratories. The study was so big that it required a steering committee of 20 scientists under the direction of NIH's Dr. Eugene Passamani. The committee included the 13 Principal Investigators from each of the clinical centers. Since researchers already knew that SK had been approved, the first phase of the TIMI trials did not include a placebo control group. The trials were strictly designed to see whether *t*-PA could dissolve blood clots better than SK. The initial trials, which began in 1984, looked only at blood-clot-dissolving abilities. They did not investigate how long the patients lived after taking either of the drugs. This turned out to be a key omission, since patients could and did die from drug side effects, such as brain hemorrhages due to blood thinning.

It was claimed that *t*-PA, which had been developed by Genentech and a Belgian scientist, would act strictly on the blood clot and not thin out the blood systemically. Unfortunately, it did just that. "It did have an effect on the blood, if you gave enough of it," testified Victor J. Marder, MD, co-chief of the Hematology Unit at the University of Rochester.[8] Marder had served on the NIH's TIMI Safety and Data Monitoring Committee until 1985, when he and most of the rest of the committee were "relieved" of their committee jobs. Marder had objected that the first phase of the TIMI study had been cut short after it showed a statistically significant advantage for *t*-PA in dissolving clots. Marder had wanted the study continued. According to the Weiss subcommittee report, "The NIH justified . . . [relieving Marder and the others] by stating that the members were overworked, since they were also serving in a similar capacity for two other studies." Marder testified: "[W]hat struck me was not that it was an unreasonable request [to stop the first phase] but that there was an urgency to the process that we were always being pressed to agree to do one thing or another and to in this case stop the study short. There was a lot of data still to be accumulated that could have been analyzed.

"There were some aspects that were a little unexpected; namely, that the bleeding was the same in both groups even though one drug was twice as effective as another in opening a vessel."

Weiss criticized the close connection between Genentech and the NIH, in particular the company's access to the Data and Safety Monitoring Committee. He asked Marder, "Am I right in the conclusion that the Genentech scientist was there at the invitation of NIH and not at the invitation of the board?"

Marder replied, "Correct."

"Drug companies other than Genentech were not represented in any of the discussion?"

"Not in person," Marder replied, "I believe they must have been contacted from time to time. I believe considerably less than Genentech was contacted."[9]

The first published report on the TIMI trials appeared in *The New England Journal of Medicine* (*NEJM*) in April 1985.[10] It simply reported the blood-clot-dissolving advantage of *t*-PA over SK. Marder testified: "We learned subsequently that even though *t*-PA was better than streptokinase in reperfusing the vessel, that the function of the heart was not really any different with the two agents. . . . We learned that a year later, and that there was no difference in mortality follow up with the two agents." However, the paper had a conclusion that, according to Marder, was not fully backed by clinical data: "[T]he conclusion of that paper was that since streptokinase reperfused only about 35 or 30 percent of the vessels, that serious consideration should be given to not using this drug at all because it seems to be so ineffective. This was without any real evidence of the study itself that there could not be clinical benefit with streptokinase."[11]

At about this same time, a group of Italian researchers known as the GISSI (Gruppo Italiano Per Lo Studio Della Streptochinais NeliInfarto Miocardio) completed their study of 12,000 patients, published in the *Lancet* in February 1986, concluding that SK was 18 percent more effective than placebo in reducing deaths in the first three weeks after treatment. A later study centered at Oxford included 20,000 patients. It showed that SK, taken with aspirin, reduced deaths by 42 percent.[12]

At the subcommittee meeting, Weiss asked Dr. Marder, "The fact that *t*-PA had a systematic effect on the blood and didn't just dissolve the clot, that was not mentioned in the article; is that correct?"

Marder replied, "That's correct."

According to the Weiss report, the TIMI researchers should have known about the systematic blood-thinning effect of *t*-PA because researchers at Temple University had already established that fact the year before. "[I]n 1984, studies comparing *t*-PA and SK found that both drugs have a systemic effect on the blood that could cause bleeding. . . . Similar results were reported in March of 1985 at the Annual Meeting of the American College of Cardiology, and during the next few years." But this result "remained unpublished by the NIH grantees until 1987 and 1988, after articles by other researchers citing bleeding problems had already been published. . . . As a result of the delay in publishing those findings, many physicians believed that *t*-PA was safer than it really was."[13]

Dr. Marder testified, "I know of one patient who bled in February 1986, and this is a very important case because the patient was not a part of the TIMI trials. This was an elderly man who had peripheral vascular disease and was having pharmacologic studies done and this particular patient received a very small dose of agent, only 15 milligrams and had a bleed, an intracranial bleed and died of the bleed."[14]

There was another key omission in the reporting of these trials. According to the Weiss report, at least 13 of the researchers owned stock in Genentech or held options to buy the stock at a discount. The report pointed out the obvious conflict of interest and illustrated how this could work: "[One researcher] for example, could have easily earned more than $250,000 exercising his stock options, if it hadn't been for the relatively negative research reports from scientists who were not funded by NIH or Genentech."[15]

It was also pointed out that the authors of the key 1985 *New England Journal of Medicine* article, which gave the first comparison between *t*-PA and SK, included "at least five researchers" who owned stock in Genentech. A subsequent article on the TIMI trials in the same journal, in 1987, indicated in a footnote that four of its coauthors held stock in Genentech.

NIH guidelines neither restricted nor required disclosure of this stock ownership.

In May 1987, an FDA panel gave preliminary approval to SK for intravenous use, but it refused to give the same initial go-ahead to *t*-PA. Genentech's stock dropped $11.25 on the Monday following the announcement. *The Wall Street Journal* denounced the FDA in an editorial titled, "Human Sacrifices." In November of the same year, the FDA reversed itself and announced at a press conference the approval of *t*-PA, asserting that, according to the Weiss report, "new data had become available." Did the pressure from *The Wall Street Journal* and elsewhere cause the FDA to change its mind on *t*-PA? According to an FDA committee, "The acrimonious debate over *t*-PA that was aired in the press did nothing to hasten its approval."[16]

In March 1990, Italian researchers conducting GISSI-2 announced results of *t*-PA research based on 20,000 patients in 13 countries: They found that SK was just as effective as *t*-PA in saving lives although *t*-PA cost 10 times as much as SK. Newspaper reports quoted U.S. cardiologists, in particular, one American cardiologist, who claimed that the Italians had not administered *t*-PA in the most effective way. The Weiss report pointed out that the article "failed to mention that, in addition to receiving research grants . . . [that cardiologist] had served as a paid consultant to Genentech for several years, and had received options to buy 6,000 shares of stock."

The Weiss report points out, "It is not possible to determine whether or not stock ownership clouded the judgments of scientists" involved in various aspects of this research. However, the bottom line on this case, according to the Weiss report was that there are "repeated examples of more positive evaluations of *t*-PA by scientists with relationships with Genentech, compared to scientists without such relationships. In many cases, the more positive evaluations were funded by NIH, and the less favorable ones were conducted by international groups."

THE CANTEKIN DISPUTE AND THE BLUESTONE CASE

Normally, cases of questionable conduct are named after the person charged with that conduct. The case to be described here is commonly referred to as the case of "Dr. Charles Bluestone and the Pittsburgh Medical School," or more briefly as the "Bluestone" case. However, in a bizarre and for science, tragic, way, this is also the "Erdem Cantekin" case. Cantekin, who left his native Turkey at the age of 21, earned his doctorate in the United States and rose to become a tenured full professor in the Department of Otolaryngology (ear, nose, and throat) at the University of Pittsburgh School of Medicine. He is the author or coauthor of more than a hundred published scientific papers, many of them coauthored with Dr. Bluestone. Until the events in this case unfolded, Cantekin had been Director of Research of the Department of Otolaryngology of Children's Hospital, Pittsburgh. Subsequently, he was fired. According to a hearing board appointed by the dean of the Medical School, he committed "serious violations of research ethics," because he "fraudulently present[ed] data" from a major study of antibiotic treatment of children's ear infections "as if he was responsible for it."[17] That he did no such thing would later become a matter of congressional record. What Cantekin actually did was to try to publish what he believed to be the truth, even though it flew in the face of established orthodoxy.

Because of his position as Director of Research for the Department of Pediatric Otolaryngology, Dr. Cantekin was also the Director of Research at the Otitis Media Research Center (OMRC), which was a unit within the Department of Pediatric Otolaryngology at Children's Hospital of Pittsburgh. The Center specializes in studying treatment for the most common childhood ear disease, otitis media. The research center was set up by the National Institutes of Health (NIH) in 1980 and had received $18.5 million in funding by 1990. The money went to specific grants for specific studies. The University of Pittsburgh staffs the Center, and its Principal Investigator is Dr. Charles D. Bluestone, a well-known ear surgeon. The NIH is not the Center's sole source of funding; by 1990 major pharmaceutical companies had also paid a total of $3.5 million for specific studies—as many as 40 of them—according to Dr. Cantekin.

The Center had been started up with a $12-million NIH grant, "to study the effectiveness of the antibiotic amoxicillin for various forms of otitis media," but according to Dr. Cantekin, then the pharmaceutical companies also began pouring in money "to determine the effectiveness of other antibiotics. In doing research for those companies, the Center would not use placebo controls and would compare the new drug against amoxicillin, the effectiveness of which had never been determined and which was in fact the subject of the federally funded research."[18]

Beecham manufactures amoxicillin, which *The Merck Manual of Diagnosis and Therapy* describes as the standard antibiotic for treatment of middle ear infections in children under eight years old. Other drug manufacturers have been eager to get in on the action. Every year, according to the Weiss committee report, "parents spend between $400 and $700 million" on antibiotics to treat their children's middle ear infections. Vast sums of money have been spent for a treatment that is unproven at best.

Dr. Bluestone did not wait for the results of the NIH study to determine if amoxicillin is effective before touting antibiotics in treating this disease. He "established himself as a relentless promoter of the use of antibiotics to treat any form of otitis media, in the lecture circuits paid by pharmaceutical companies," according to Dr. Cantekin.[19] Cantekin added, "While the Center was in the process of performing the NIH-funded study of the effectiveness of amoxicillin in the treatment of both acute and secretory otitis media, this primary investigator was publicly advocating the use of amoxicillin for these conditions. In a paper published in *Acta Otolaryngol* (Stockholm) . . . he opined that amoxicillin produced a short term resolution of otitis media in 50% of cases treated. . . . In his pro-antibiotic lectures, this investigator footnoted the NIH-funded amoxicillin studies underway, thereby implying that such studies, the results of which were not yet known or published, in some manner supported his pro-antibiotic position. This primary investigator spent substantial time traveling at the expense of the pharmaceutical industry to advocate use of their products, collecting honoraria in the process."

Cantekin felt it constituted "a conflict of interest" for a scientist involved in federally-funded testing of a drug of a private company to accept honoraria from the producers of those drugs. In April 1987, Cantekin put his objections in writing, in a letter to the dean of the Medical School. When he received no reply, he wrote the NIH and FDA to file an official complaint.

The NIH investigated Cantekin's charges concerning the honoraria received by Dr. Bluestone from pharmaceutical companies between 1983 and 1988 and found that Bluestone had indeed received noticeable sums in honoraria and travel expenses from such companies. The amounts varied from year to year: The least came to $15,050, in 1983; the most totaled $60,738.32, in 1985; and the combined amount added up to $262,067.32. The money from pharmaceutical companies overwhelmingly made up the bulk of the honoraria he received. The honoraria and travel expenses from sources other than pharmaceutical companies during the same period came to only $33,550.[20]

"Honoraria are a big problem," testified the editor of the *Journal of the American Medical Association,* Dr. George D. Lundberg. He added, "Obviously the word is a euphemism. Honoraria are hardly gifts. They're almost fees."[21]

Asked to comment on the amount of money Bluestone received in honoraria from pharmaceutical companies, the dean of the University of Pittsburgh Medical School, Dr. George M. Bernier, Jr., testified, "I would say it's on the high side."[22]

Dr. Bluestone did not testify at the Weiss hearings. But portions of an interview he gave were on the French television network, TF-1. He said, "If I wanted to make money, I certainly wouldn't go on the stump and spend two days and get five hundred or one thousand dollars—because a surgeon makes that in five minutes. . . . If my colleagues who are in this country and abroad felt that I was biased by pharmaceutical companies, I wouldn't be invited because that means that I'm selling products made by that company. They believe that I'm unbiased. That's why they continue to invite me to speak."

In a June 26, 1989, unpublished statement submitted to the Weiss subcommittee, Bluestone stated that Cantekin's allegations "are meritless." He argued that "it is important to educate physicians" as well as keep them informed of research results. Furthermore, he stated that industry funding for travel and honoraria "is not unique to me." In an earlier unpublished statement submitted for Weiss's September 29, 1988, hearing, he said that one of the Center's objectives is specifically to "disseminate information" from their research. He further argued that as an ear surgeon, his income "is primarily derived from surgical fees," which would be reduced if the amoxicillin treatment works.

The NIH wrote, "Even though the University quite clearly encourages its faculty to interact with and seek funding from the pharmaceutical industry and Dr. Bluestone has done so within the university guidelines, there is concern that the situation involving Dr. Bluestone gives the appearance of a conflict-of-interest."[23] However, the same report cleared Bluestone of an actual conflict of interest in that the analysis of his studies didn't appear to show a systematic bias in favor of the efficacy of antibiotics. Under prodding from the Weiss subcommittee, the NIH did a subsequent analysis and came to a different conclusion—that Bluestone and his colleagues had not evaluated the "biological significance of their data," and as a result their reports were "less than objective."[24]

In addition to the conflict of interest, Dr. Cantekin objected to the money the Center collected from pharmaceutical companies, testifying, "In June 1983, I met with my department chairman and informed him that I had grave doubts about the scientific validity of research commissioned by and funded by pharmaceutical companies seeking to prove the effectiveness of their antibiotics. I argued that the Center should disassociate itself from industry-sponsored antibiotic effectiveness research, and should perform studies only when funded by neutral sources. My department chairman replied that I was free as a matter of personal choice to disassociate myself from industry-sponsored work, but that he could not require a primary investigator of the Center to refuse industry funding."[25]

In a unpublished statement for the record of Weiss's 1989 hearing, the department chairman stated that he had "no record or recollection" of the conversation Cantekin described.

Cantekin's Rejection of Industry-Sponsored Research

Dr. Cantekin, as Director of Research for the Center, did refuse to participate in any further industry-sponsored research because he felt such funding biased the findings. He explained:

"For example, the Center performed a study to determine whether a certain antibiotic, Cefaclor, manufactured by Eli Lilly & Co., was effective in the treatment of otitis media. This project, funded by Eli Lilly & Co., produced two published papers. The first reported that Cefaclor was as effective as amoxicillin in the treatment of acute otitis media. . . . The second paper reported that Cefaclor should replace amoxicillin in the treatment of acute otitis media in children with frequent attacks. During the preparation of this second manuscript, I dissented on the basis that the data generated did not support such a conclusion. I also felt that the source of the funding for the research, Eli Lilly & Co., affected scientific methodology and reasoning. This second paper was also published."

Cantekin noted that the research reported in the second Cefaclor paper—that Cefaclor was superior to amoxicillin in the treatment of acute otitis media—was not supported by three other studies on Cefaclor's effectiveness at the University of Minnesota, Case Western Reserve University, and the University of Texas.

In his June 26, 1989, unpublished statement to the Weiss subcommittee, Dr. Bluestone stated that another study conducted by the Center [AB-OME-II, to be discussed below] produced an outcome which is "*adverse* to Eli Lilly's product!"

Another objection Cantekin had to industry-sponsored research at the Center was that it delayed "the completion and dissemination of NIH-funded research. I think that is the most important factor of this conflict of interest issue because we have completed six clinical trials during the last few years and only one of that six is published, due to the fact that manpower is used to disseminate the pharmaceutical company research. . . . The pharmaceutical companies don't have indefinite time limits like the NIH does. They want their answers as fast as possible."[26]

What kinds of research are going unpublished, as a result of industry-sponsored research? Cantekin explained: "We have a clinical trial finished in 1984 which shows a common surgical procedure used in ear disease management, myringotomy—lancing the eardrum, is a totally unnecessary surgical procedure. This is not yet published in a refereed journal and is not available to the public at large."

The then director of the NIH, James Wyngaarden, MD, testified, "[T]he decision as to when research is mature and thoroughly analyzed and digested and ready for publication is in the hands of the investigator."[27] In his June 26, 1989, statement to the Weiss subcommittee, Dr. Bluestone stated that the Center did "not compromise" either the quality of its analysis of data or of its written reports "in the interests of speed," but is in fact committed to "reasonable and appropriate" timing of its reports to the NIH.

When the NIH looked into the allegations of unpublished studies, however, it found that of 12 studies that had not yet been published, two studies involving clinical trials of children "should have been completed and published by now."[28] In effect, the NIH not only is not getting all of its own studies published in a timely manner but is also subsidizing industry-sponsored studies. A number of these studies, on publication, have given the NIH credit as a sponsor of the research, along with the pharmaceutical company. "The purpose of that," according to Congressman Weiss "is to put the prestige of the NIH behind a study which, in fact, has been funded by the pharmaceutical company itself."

Although Dean Bernier thought putting the NIH stamp on the studies was acceptable, he could not explain why the Center had not informed the NIH that the pharmaceutical companies had paid for the studies on the effectiveness of antibiotics in treating otitis media.

The Unmentioned $3.5 Million

One of Cantekin's charges against Bluestone was that he had not informed the NIH of the extent of industry involvement in research the NIH was paying for. Cantekin testified in 1988: "The pharmaceutical industry was, of course, fully aware that the Center had the NIH-funded project underway to determine the effectiveness of amoxicillin in the treatment of otitis media. The pharmaceutical industry has a large financial stake in avoiding a conclusion that amoxicillin is not effective. . . . The Center evidently understood the impropriety involved in permitting the pharmaceutical industry to funnel more than $1.6 million into the Center while the Center was performing NIH-funded research which could have a large impact on the industry. It was doubtless that awareness which caused the Center to affirmatively conceal from the NIH the extent of its reliance on the pharmaceutical industry."[29]

The NIH investigated this charge and said in a June 1, 1989, memo: "We found merit to the allegation that the [research center] has not generally disclosed to NIH the extent of its industry-sponsored research. Dr. Bluestone maintains that NIH was fully informed of his outside funding through citations in progress reports, published studies, etc., although not necessarily in the appropriate area on the grant application form. In a June 1987 letter to NIH, Dr. Bluestone stated that 'we fully reported the progress and publication status of all of our studies related to otitis media research which NINCDS-funded [an agency within the NIH] or funded by other sources, e.g., pharmaceutical companies.'

"We reviewed the documents Dr. Bluestone submitted in defense of this contention and found that, although a few studies funded by pharmaceutical companies were cited specifically in submissions to NIH (although not in the required area of the application form), no dollar amounts of funding were given and in most cases it was mentioned only that other projects were ongoing for which funding had been received from 'other federal, corporate and private sources.'"[30]

In his June 26, 1989, statement to the Weiss subcommittee, Dr. Bluestone said the NIH was "incorrect" in its characterization of his reporting of pharmaceutical company funding. He stated that "we reported *most* (not a *few*)" of the pharmaceutical company funded studies.

According to the Weiss subcommittee report, "Dr. Bluestone later admitted [the amount of funding] totalled $3.5 million. . . . Five of the funding sources were cited in 1986 or later, after Dr. Cantekin had begun to make allegations against Dr. Bluestone."

After Cantekin complained to the NIH, the dean of the Medical School appointed an ad hoc committee, the "Medsger committee," to look into the charges. The Medsger committee reported back to the dean that Cantekin's charges "did not have substance or validity."[31] Dean Bernier then wrote to the NIH and informed the Deputy Director for Extramural Research that the University of Pittsburgh School of Medicine did "not intend to pursue the allegations further."[32]

On June 23, 1987, one day after the Medsger report was submitted and the Medical School decided not to pursue the matter, Dr. Bluestone wrote the NIH that while the Center was going through its documents to show their full disclosure, they did discover an "omission" concerning one study sponsored by the NIH. This study compared the efficacy of amoxicillin to both a placebo and a decongestant-antihistamine. According to the letter, the study was redesigned in 1984 to compare both amoxicillin and a placebo with two other antimicrobial agents. This required, the letter explained, "a cost sharing arrangement" that "was not explained fully" in a February 1, 1985, grant renewal application.[33]

Bluestone's letter supports at least in part Cantekin's allegation that full disclosure was not made even though the charge had been dismissed by the Medsger committee.

The NIH excused the omission because they said their forms calling for Bluestone to list the outside funding were not clear. The form stated: "List *all research support* for each individual including requests now being considered, as well as any proposals being planned, regardless of relevance to this application. Include *all current or pending* contracts, fellowship awards, research career programs awards, training grants, *regardless of source of support.* List grant number, total funds for the entire project period, estimated percentage of effort the individual devotes to the project, and *the source* of the support [emphasis added]."

Although the Weiss subcommittee report seemed to think this was clear enough, the NIH did not agree. The Weiss report stated, "NIH apparently considered these instructions vague because they did not specifically define the requirement of listing 'all' sources of funding as including 'non-Federal' sources of funding."[34]

The Amoxicillin Studies

Cantekin's primary allegation against Bluestone was that Bluestone's paper on the big amoxicillin study showed an effectiveness that was not true and was obtained only by changing the diagnostic criteria for the disease: "The clinical phase of the NIH-funded study to determine the effectiveness of amoxicillin in the treatment of persistent (secretory) otitis media was completed in 1984. Draft reports were circulated within the Center, arguing that amoxicillin was shown by the data to be effective in the treatment of secretory otitis media. I participated in the discussion of these papers and took the position that the data did not show effectiveness. I declined to be a co-author and bowed out of the manuscript on the thirteenth draft in September, 1985. . . .

"In February, 1987, the Center's amoxicillin manuscript was published in the *NEJM* [*New England Journal of Medicine*]. The paper concluded that amoxicillin was effective in the treatment of secretory otitis media. The authors of that paper, however, had changed the definition of the disease [e.g. diagnostic criteria for the disease] from the definition contained in the original grant application to NIH. Under the original definition, no argument could be made that amoxicillin had any efficacy in the treatment of secretory otitis media. The authors of the Center's amoxicillin paper did not, however, explain or justify the changed definition of the disease and did not have approval of NIH for the change."[35]

According to the Weiss report: "Dr. Bluestone and his colleagues did not diagnose children's ear infections as they had said they would in their NIH grant proposal, and instead analyzed it in a way that made the drug under study look more efficacious . . . Bluestone's change in the diagnostic algorithm was an objective fact that was incorrectly denied by the Medical School investigative committee, based on Dr. Bluestone's denial (which he later retracted)."

According to Cantekin's lawyer, the effect of the change is "to overstate the number of 4-week cures in the treatment groups by at least 10%."[36]

Part of the controversy surrounding the change in definition involves the reliance in the Bluestone paper on otoscopy rather than tympanometry to diagnose the illness. An otoscope is an instrument that looks like a flashlight with the light coming at the end of the "pipe" but at a right angle to it, and with a cone (speculum) at the top. By inserting the cone into the ear, it is possible to peer directly at the eardrum. Usually, an air bulb is attached to put pressure on the eardrum so the doctor can observe it move. The main point of otoscopy is that the doctor makes a judgment simply by visually examining the eardrum. Tympanometry uses a sonar mechanism that measures the echo and can detect fluids not visible by looking into the ear. An outside statistical analyst contacted by Dean Bernier and used by the hearing board commented on otoscopy in a November 28, 1988, letter to Brian V. Jegasothy, MD, head of the hearing board: "[It] is a subjective measure which is one component of the primary endpoint, and as it turns out is the one which most favors the experimental treatments. . . . It is entirely possible that aspects of a patient's record could have given clues regarding treatment assignment, and that conscious or unconscious bias could have crept into the assignment. However, there is no way of knowing whether this was indeed the case without an independent review (which is how such trials should probably be designed in the future). It may be that replication of the trial is the only way to resolve the current dispute."

Cantekin coauthored a paper on peer review that was published by *JAMA;* he detailed the otoscopy issue in a footnote and contended that there was ". . . strong statistical evidence of systematic bias in the otoscopic measurements and that analysis should be done using tympanometry. Using tympanometry, it is indisputable that the . . . data show no drug efficacy. The OMRC used otoscopic measurements and scored ears as having no effusion on the basis of a negative otoscopy reading, whereas the corresponding tympanometry measurement suggested a high probability of effusion." In other words the sonar echo showed the likelihood of fluid that the doctors could not see.

Another part of the controversy was at what point to look at each child to decide if he or she was "cured." Cantekin noted in *JAMA* "Whether the end point for analysis of antibiotic efficacy should be 4 weeks or whether it should be based on 2-week, 4-week, and 8-week data [Cantekin], argued that there was a strong recurrence pattern in the antibiotic-treated group that made it improper to limit the analysis to a single time point of 4 weeks or less."[37]

Congressman Weiss asked then NIH Director Wyngaarden, "In this instance, if the diagnosis, which is described in detail in the grant proposal, leads to a finding that a drug was not significantly better than a placebo or sugar pill, and the diagnosis used in the study resulted in a finding that the drug was effective, wouldn't this be something that should be checked out by NIH, the granting agency?"

Wyngaarden replied, "If it really changed the criteria of the study and thereby allowed a result to be positive whereas otherwise it would be negative, I think that's a very major change. . . ."[38]

Why Was the Study Cut Back?

A third issue in the dispute concerned a change in the total number of subjects in the study—the number of subjects was cut in half. In the *JAMA* article, Cantekin explained: "In 1984, the AB-OME-I clinical trial was terminated by the OMRC with half the target sample size (518 patients). A new clinical trail, AB-OME-II, was then initiated that continued the AB-OME-I protocol but added two additional antimicrobials (cefaclor [Ceclor] and Pediazole). Funding from two pharmaceutical companies was obtained for AB-OME-II without the knowledge of the NIH, which continued to fund AB-OME-II under the NIH program project grant."

Cantekin charged that the funding from the pharmaceutical companies was related to financial difficulties the Center was having. He wrote William Raub, of the NIH on June 27, 1989: "This early termination was not based on any sound statistical criterion and was prompted by the desire of OMRC to obtain private pharmaceutical funding to evaluate Pediazole and Ceclor, two additional antibiotics not part of the NS-16337 grant. At the time of bringing private pharmaceutical support into the OME trials, the OMRC had a deficit in the range of $328,000 and it was desperate to obtain an infuse of money from private pharmaceutical companies."

Bluestone acknowledged the deficit in a June 23, 1987, letter to the NIH in which he stated that Children's Hospital had informed him of a deficit in their research account that reached $328,000 by January 31, 1984. The reason for the deficit, wrote Bluestone, was that the acute otitis media study had been conducted "in two sites," and there was not enough money to cover the clinical trials. So, in 1984, the Center started paying Children's Hospital from "non-NIH funds."[39]

Cantekin explained in a telephone interview with me on May 7, 1991, "The $328,000 deficit came about because they couldn't get enough patients through Children's Hospital of Pittsburgh because the doctors there wouldn't refer enough of them. So they were paying roughly $1,000 per patient to two private pediatric clinics, one in the suburbs, one in town, to enroll the patients in the clinical trials, administer the antibiotic or placebo and then follow up the patient. . . . the real number paid to these pediatric private practices is much larger than $320,000 since they were also used in the many studies paid for entirely by pharmaceutical companies. My guess is that these pediatric practice groups received between $1.5 million and $2 million over several years. They got about half the money the drug companies paid out to the OMRC."

In his 1988 statement to the Weiss subcommittee, Bluestone strongly insisted that the $1,000 per child was not paid to "enroll" the children in the study but was "paid to the physician for services rendered in treating the child." In his June 6, 1989, statement to the Weiss subcommittee, Dr. Bluestone stated that "he takes strong exception" to the suggestion that there is "something improper" in shortening the study. He believed researchers should be allowed the flexibility and freedom to make a number of changes, one of which is to "shorten trials."

Congressman Weiss asked NIH Director Wyngaarden, "Are you satisfied with the fact that an applicant makes a grant application, saying that he's going to study 680 children and ends up with 164?"

Wyngaarden replied, "I don't know the circumstances. If the answer is clear-cut, yes. For example, if 164 were so clearly positive or so clearly negative, we've shortened many a clinical trial on that basis."[40]

But the results were *not* clear-cut. Dr. Cantekin noted that the *NEJM* insisted that the authors tone down the conclusion of the OMRC paper published in the journal. Rather than stating that amoxicillin was effective and should be used in the treatment of secretory otitis media in children, the conclusion was amended to say that the drug was marginally effective.[41]

The NIH's Office of Scientific Integrity (OSI) investigation into the case in 1990 came to a similar conclusion: ". . . since 'cure' rates are relatively low (about 30% or less) for the percentage of patients effusion free at four weeks by whatever criterion is utilized and since it is agreed that the recurrence rates are relatively high subsequent to four weeks, it is evident that not many patients have long-term benefit from the administration of antibiotics. Thus, it would have been desirable for the authors to have discussed the biological significance of their findings in greater depth."[42]

The Exceptionally Unusual Placebo Cure Rate

Cutting the sample size in half may have resulted in an advantage to the drug companies helping to fund AB-OME-II, another antibiotic effectiveness study for otitis media. The percentage of cures in the placebo group of the AB-OME-I clinical trial was *half* that found in other major studies. A surprisingly low cure rate for the placebo-treated group might have made an otherwise unsurprising cure rate for the experimental group (amoxicillin-treated) look statistically significant. Cantekin listed the cure rates for three OMRC studies and one Danish study done since 1978. The groups receiving the experimental drugs had roughly the same cure rate. The placebo-treated control groups were also roughly the same, except for one, which varied widely:

D&A Study, 1978–1981	.240
Thomsen et al., 1984–1986	.310
AB-OME-II, 1984–1987	.267
AB-OME-I, 1981–1984	.141

AB-OME-I produced a placebo cure rate far below that found in the other studies. According to Cantekin, this should not be the case. If the control groups have a statistically significant number of patients, the control groups should all produce more or less the same spontaneous cure rate—unless the nature of the disease changes, or the diagnostic procedure changes, or the criteria for identifying the disease changes, or the population being studied changes.

Cantekin pointed out to the NIH that the conclusion reached by the "Mandel et al." paper, in the *NEJM*—that amoxicillin was 17% more effective than placebo in treating OME in children was based on the anomalous placebo cure rate in AB-OME-I. Cantekin concluded that if the highly unusual placebo cure rate were thrown out as anomalous or combined with the placebo cure rates for the Center studies listed earlier, a valid argument could not be made for the effectiveness of amoxicillin. This would be the case, he said, regardless of the "analytical technique" used, including the one in "Mandel et al.," which Cantekin characterized as "dubious."[43]

The NIH's Office of Scientific Integrity (OSI) investigation into the matter did not discuss Dr. Cantekin's charges regarding the placebo.[44]

Cantekin was especially upset with this anomalous placebo cure rate because at the time the "official" paper ("Mandel et al.") was submitted for publication to the *NEJM,* the Center *already knew* the placebo cure rate for the second half of the same study, AB-OME-II, was substantially higher: "Dr. Bluestone and his colleagues proceeded to publish Mandel et al. in the *NEJM* (Feb., 1987) *with no reference whatever to either the AB-OME-II data or the very different placebo cure rate in the 1983 DA study* [emphasis in original]."[45]

The OSI Report stated: "[C]omparisons of the active drug with placebo had not been featured as prominently as should have been the case, possibly because such comparisons would have weakened the case for antibiotic efficacy that apparently was made in the OME-I trial."

Dr. Cantekin elaborated on the point in his letter to the NIH, stating that the panel had not commented at all on this matter, but that it was, in Cantekin's view, in the most extreme category of seriousness for researchers to permit their paper to be published with conclusions they knew were contradicted by their own later experiments. He asked, "How can the panel pass this point by without comment?"[46]

Why did the placebo cure rate vary? Cantekin noted some prognostic factors of the disease that may make it more resistant to cure, such as whether the child has the disease in both ears (bilateral) rather than just one. Random assignment should distribute the various prognostic factors roughly equally among the groups. "In fact, wrote Cantekin to the NIH, "the data from the completed AB-OME-I study indicate that the placebo-treated group has a higher incidence of bilateral disease by a substantial amount." This factor would result in a *lower* spontaneous cure rate in the placebo control group. Curiously, a 1983 annual report submitted to the NIH, when only about 60% of the 518 subjects had been tested, showed that at this point in the test, the situation was the opposite; one of the *amoxicillin groups* contained a larger number of patients with otitis media in both ears (which would have skewed the results against amoxicillin). By the time the experiment was completed, however, it was the placebo group that was significantly skewed, with an abundance of children having otitis media in both ears. This group's cure rate, as noted earlier, was anomalously low (14%). Cantekin thought it unlikely that the final 40% of the subjects had been randomized in a way that reversed the already existing skewing. This would require the randomization to assign the control group a sufficient number of patients with bad prognostic factors.

Burying the Results of the Control Group

The report given by the Center to the NIH on AB-OME-II placed the data regarding the three antibiotic-treated experimental groups in a different table from the data on the placebo group. In corporate accounting this is sometimes called "burying" the results from the watchful eyes of investors. The auditors for the

NIH, the Panel of Scientists, did in fact find "comparisons of the active drug with placebo had not been featured as prominently as should have been the case."

To compare the placebo control group with the three experimental (antibiotics) groups, the OSI investigators had to look at two different tables. Normally in science the experimental and control groups are reported on the same table because it makes comparison easier.

Nevertheless, after the OSI's Panel of Scientists had compared the two tables, they wrote, "The data at four weeks provided no real evidence that any of the antibiotic therapies was superior to placebo."[47]

This is precisely what Cantekin had been protesting all along. One of the tables in the Center's report compared the three antibiotics, with the claim that amoxicillin was the "control" in that case. According to the OSI's report, neither of the two new antibiotics was better than the old standby, amoxicillin. But was amoxicillin effective? Table 4 showed that it was not, after four weeks. The text in the Center's report did not elaborate on this fact. The OSI report stated, "Emphasis was given to significantly superior results for A over P at two weeks ($p = .015$) but not at four weeks ($p = .80$), the endpoint used in the OME-I trial."

Cantekin, writing about the Center's report to the NIH, said that the way the Center had reported the results was not "acceptable scientific conduct." Their report had, in effect, stated that the standard antibiotic had produced a certain cure rate, and that a challenger had not done any better. What their report had not "simultaneously" disclosed, said Cantekin, was that the standard antibiotic worked no better than did the placebo used in the same study. Cantekin argued that the way the Center's report was written conceals that AB-OME-II contradicts AB-OME-I. He condemned this way of presenting the results.

In his June 26, 1989, statement to the Weiss subcommittee, Dr. Bluestone defended the presentation of the material, arguing that "the purpose" of the study was "not to compare" alternative antibiotics with a placebo but with amoxicillin. He also argued that by combining the amoxicillin versus placebo data from AB-OME-I and AB-OME-II, amoxicillin was shown to be effective at four weeks.

After looking over the AB-OME-I and AB-OME-II studies, the OSI panel wrote, "It seems evident to the OSI panel that these data do not provide support for the long-term effectiveness of *any* of the antibiotic drugs, relative to placebo, through a period of 16 weeks."

What is the reaction of the scientific community to the publication of the results of AB-OME-II? Four years after the AB-OME-II study was completed, the results apparently failed to meet NIH Director Wyngaarden's "mature and thoroughly analyzed and digested and ready for publication" test. Cantekin wrote, in an August 13, 1990, letter to Suzanne Hadley of the Office of Scientific Integrity, that the AB-OME-II study, which "contradicts" the AB-OME-I study reported in "Mandel et al.," had not been published.

There has since been independent evidence that the efficacy of antibiotics in treating otitis media is negligible. The *British Medical Journal* published the results of a study of more than 3,600 children with ear infections, showing that antibiotics were not effective for treatment and that rates of recovery were *higher* for patients who did not receive antibiotics. The study concluded, "Antibiotic treatment did not improve the rate of recovery of patients in this study."[48]

The Weiss report said of the British report, "Similar evidence of the ineffectiveness of antibiotics would have been available to physicians and the public several years ago, if the Medical School had not prevented Dr. Cantekin from publishing them."

This 1990 British study is interesting in part because of the sworn testimony of then NIH Director James Wyngaarden, MD, given in 1989, "In my mind, some of the rather unusual steps that Dr. Cantekin has taken might be viewed differently if his view is sustained scientifically."[49]

Cantekin has not had access to the Center's final report on the AB-OME-II study. Both the Center and the NIH have refused to give him a copy. He based his comments on the draft of the OSI 1990 report sent him by the NIH. Here's what Joanne Belk, NIH's Freedom of Information Act officer, wrote Cantekin on July 11, 1989, after he requested Bluestone's report on AB-OME-II: "In accordance with our Freedom of Information procedures, we notified Dr. Bluestone that his report had been requested. His attorney, Wilbur McCoy Otto, responded and pointed out that much of the content of the report was patentable. That assertion has been confirmed by the NIH Patent Attorney. Thus, I must deny access to the report under the provisions of 5 U.S.C. 552(b)(4) of the Act and 45 CFR 5.65 of the Regulation. Exemption 4 protects commercial or financial information obtained from a person which is privileged or confidential."

In Cantekin's reply back to the NIH he wrote that the basis for the proprietary claim is that the OMRC received funding for the study from both the NIH and pharmaceutical companies. The money from the private companies, which Cantekin characterized as "unauthorized and undisclosed," turns this study, which was also funded by the federal government, into a "commercial trade secret."[50]

Publish and Perish

In the midst of the controversy Dr. Cantekin testified: "It is obvious that the Center is going to be very resistant to concluding that amoxicillin itself is not effective when it has been publishing papers for years in which it compared new drugs against amoxicillin on the assumption that amoxicillin was the standard drug in the treatment of otitis media."[51]

Cantekin ceased participation in the "Mandel et al." manuscript, in September 1985 and began to object in writing, in the form of memoranda to the other researchers, including the principal investigator. "The Center's primary investigator and the other researchers ignored my criticism," says Cantekin.

At this point, he teamed up with Timothy W. McGuire, PhD, a well-known expert in statistical analysis, from Carnegie-Mellon University, and they wrote their own paper. Referred to here as "Cantekin et al.," the paper stated, "[A]moxicillin was not effective in the treatment of secretory otitis media." Cantekin testified that although he "originally listed the primary investigators of the Center as authors of this paper, they declined."[52]

Cantekin tried to settle the dispute before submitting his paper, but the Center sent its paper into the *NEJM* anyway, in May, 1986.

Cantekin then met with his department chairman and told him he planned to submit his own paper to the *NEJM:* "He said he had no objection and would like a copy. . . . It was the feeling of my chairman that the *NEJM* would have the two competing views . . . reviewed by external referees and that out of that review process would come an opinion of which paper had scientific validity."

A month after the "Mandel et al." paper was submitted, Cantekin sent his manuscript off to the *NEJM.* The Medical School's Executive Hearing Board, headed by Brian V. Jegasothy, MD, later concluded that Cantekin "wrote a parallel publication, fraudulently presenting the data as if he was responsible for it."[53] Cantekin's lawyer wrote the University: "Dr. Cantekin *never* 'fraudulently presented' anything to anyone in connection with the AB-OME-I manuscripts. In his cover letter to the *NEJM* in 1986, Dr. Cantekin specifically adverted to the earlier submitted manuscript of the OMRC, "Mandel et al.," and indicated that his results were contrary."[54]

On June 30, 1986, Cantekin wrote to Arnold S. Relman, editor of the *NEJM,* that he had an "obligation to submit" his manuscript for publication. He specifically referenced the manuscript sent one month earlier, "Mandel et al.," and pointed out that his own manuscript shows "considerably more ambiguous results," even though it is based on the same data. He noted that more than once he had "alerted" the authors of "Mandel et al." to the matters discussed in his own manuscript, but they had chosen to dismiss his concerns. Then he wrote a key phrase, "as a co-investigator (Dr. Cantekin) we had no other option but prepare this manuscript."

He also pointed out that he had submitted copies of the manuscript to the authors of "Mandel et al.," and that one, Dr. Rockette, had made some suggestions that he intended to incorporate but would not significantly affect the results in the paper.

Cantekin did not pass himself off as the principal investigator, but clearly identified himself "as a co-investigator." Admittedly, his letter does imply that he felt some responsibility that a scientific paper report the truth, but every scientist should feel such responsibility. Cantekin's letter emphasized that neither he nor his coauthors had any appearance of a conflict of interest of any kind. They were not consultants, nor did they have any other sort of arrangement with pharmaceutical companies. There is no evidence that the issue of conflict of interest was taken seriously by the *NEJM.* Cantekin would later write in *The Journal of the American Medical Association,* that when the *NEJM* accepted "Mandel et al.," the journal asked for a conflict of interest disclosure from the OMRC. But, wrote Cantekin, the OMRC did not inform the *NEJM,* just as it had not previously informed the NIH, that the OMRC "had received millions of dollars in funding from the manufacturers of antibiotics, including manufacturers of amoxicillin."[55]

Dr. Bluestone has persistently maintained that he met all NIH filing requirements, not by stating specific "dollar amounts," but by "extensively" describing non-NIH funded studies. He has stated that the OMRC has "never sought to conceal" that it receives funding from pharmaceutical companies, and that the receipt of this funding does not constitute a conflict of interest.

The Executive Hearing Board at the University of Pittsburgh decided that Cantekin tried to "derail the publications of his colleagues."[56] The opposite is the case—officials of the Medical School tried to and succeeded in derailing Cantekin's manuscript. In reply to an inquiry from the editor of the *NEJM* to Dr. Bluestone, the medical director of Children's Hospital, its president, and the chairman of Cantekin's department wrote the editor. Their letter stated that Children's Hospital "owned" the complete set of data on which the papers were based and that Bluestone, the principal investigator, had the "delegated responsibility" for the final scientific report which would result. They said that Cantekin "acted on his own," and they underlined that "at no time" did Cantekin either ask for or get from Bluestone or the department chairman permission to write a paper using the data. They had not even given Cantekin permission to use the data at all, or to have the data statistically manipulated by someone outside the university or the hospital. The letter stated that what Cantekin did was "unethical, improper and a source of grave academic concern." The letter endorsed "Mandel et al." as the "only authorized" scientific paper concerning the study and stressed that Cantekin's actions were taken "on his own initiative" and lacked "authority."[57]

The letter ended the *NEJM*'s consideration of "Cantekin et al." The journal returned the manuscript to Cantekin with a letter, dated August 6, 1986. The editor wrote that officials of Children's Hospital and the university had informed him that Bluestone was the principal investigator. They had also stated, said the editor, that Cantekin's actions had been "improper"; he had both appropriated the data and submitted his own interpretation. The editor referenced an earlier letter he had written Cantekin and repeated that in his view the decision as to "how and when" a scientific paper should be sent for publication was the exclusive "right" of the principal investigator.

None of the journal's reviewers considering "Mandel et al." ever saw Cantekin's version. Nor did the *NEJM* invite Cantekin to submit a letter to the editor summarizing his disagreements that could have been published simultaneously with "Mandel et al." Cantekin later wrote in *JAMA:* "It is a breach of trust, we think, for the editor of the *NEJM* to have asked the external reviewers to comment only on the principal investigator's manuscript without having available the views of the dissenting coresearcher. Similarly, the *NEJM,* in publishing the manuscript of the principal investigator, should be obligated to inform its readership of the existence of dissent and of the arguments of the dissenter."[58]

In a subsequent exchange of letters, the editor of the *NEJM* informed Dr. Cantekin's attorney that the sort of submission Cantekin had made could result in "chaos" and the "orderly business of science" could not be carried out. The editor wrote, "The important question . . . is not whose interpretation is correct . . . but rather who has the right to publish the data first and make the first interpretation."[59] Cantekin's lawyer wrote back: "Let me reword your statement just slightly to demonstrate its astonishing absurdity: 'The important question . . . is not whether Galileo or the Pope is correct with respect to whether the sun revolves around the earth or vice versa, but rather, who as between the Pope and Galileo, has the right to publish his opinion. . .'"

How about a Letter to the Editor?

Not everyone agreed with the University of Pittsburgh Medical School or the *NEJM* on this matter. Former National Science Foundation (NSF) Director Erich Bloch testified: "[J]ust on the surface of it, that doesn't sound—doesn't sound right to me. Anybody can publish anything, especially if he has done that particular work and if it leads to different interpretation or different—or if he is in disagreement with the data itself, then I think he should publish in one form or another, either by writing a letter to an editor of a particular journal that is appropriate, or writing his own paper."[60]

The head of the NIH, which funded the study, however, firmly stated that the most important factor was the orderly process of science, including who the boss is. He testified: "If this had happened in my laboratory when I was running a laboratory, I would have been incensed. I clearly would have been incensed. The award was to Dr. Bluestone. . . . in the tradition of science, if the award is to a given investigator, that investigator is the one who is responsible to us through the institution for the conduct of that research, and with that goes an assumption that that investigator is the controlling director of that research, and has the final authority to decide what should and should not be published. And I don't think that I know of any other instance in which someone who was working under a principal investigator has even attempted to publish the data on his own or her own."[61]

Wyngaarden added that there were instances where investigators had differing points of view. He stated that the "usual mechanism" for making such viewpoints public was for the collaborator or coprincipal investigator to follow up the main paper with a letter to the editor. The dean of the Medical School, George M. Bernier, Jr., MD, stated that he also had "encouraged" Dr. Cantekin to send a letter to the editor. Still, the letter from the three officials of the University of Pittsburgh Medical School to the *NEJM* claimed ownership of the "Mandel et al." data. In a sense, they closed the door to the possibility of a letter to the editor by essentially saying that Cantekin should not be taken seriously.

Cantekin himself stated that there is no standard way to publish dissent, but one clearly important element is how extensive the dissent is. If it concerns a detail of a paper, then a letter to the editor, which generally is restricted to 500 words, would be fine. But if the dissent concerns issues of bias and possible undisclosed financial ties to an industry affected by the research, a letter would not be enough. Cantekin points out that a panel criticized him for not "cramming" his dissent into 500 words, but the panel itself devoted 69 pages to the case.[62]

The actions of the *NEJM* with regard to a follow-up publication are somewhat puzzling. Arnold S. Relman, MD, the editor of the journal wrote Cantekin's lawyer, Robert L. Potter, on January 6, 1987, that Cantekin was "entitled" to state his dissent, but that unless Cantekin could convince the principal investigator to change his own paper, the "proper time" for Cantekin to publish his dissent would be after the publication of the principal investigator.

Cantekin did what Editor Relman suggested. On April 12, 1988, Cantekin submitted his dissent to the *NEJM* in the form of a six-page letter to the editor. On May 24, 1988, Relman wrote him back that the *NEJM* could not "use" his letter in their section on correspondence because it was more than 400 words long.

Observing "Tradition"

With respect to the idea that the University of Pittsburgh School of Medicine "owned" the NIH-funded data and Bluestone had some sort of exclusive right to it, the Dean's Executive Review Board, the Jegasothy committee stated that no "clear and definitive" guidelines exist concerning data ownership from a study funded by the federal government. But "by tradition," both data and first publication rights are held by the principal investigator. After he or she has published, the data, which is then "public property," can be reanalyzed or reinterpreted by others.[63]

Cantekin's lawyer disagreed: "There was no evidence whatever placed before the Hearing Board with respect to the existence of any such 'tradition.' We are unaware of any document, other than the Hearing Board Report, which argues the existence of any such 'tradition.' . . . No rational argument could be advanced that it furthers scientific truth or scientific method to prevent the reviewers at the journal to which the principal investigator has submitted his paper from reading the dissenting views of a co-investigator on the same subject.

"The Hearing Board at no point offers any word or explanation with respect to why such a 'tradition' should exist or what benefit such a 'tradition' offers to science. . . . In this case, 'tradition' offends the search for scientific truth."[64]

Cantekin added, in his *JAMA* article, "It is hard to imagine public acceptance in the United States of a theory of private ownership of data generated by publicly funded studies that could be used to prohibit dissent."

Editor Relman, of the *NEJM,* agreed with the university and wrote Cantekin that the notion of a co-principal investigator submitting on his own would result in a "chaotic welter of conflicting and uninterpretable claims."

George D. Lundberg, MD, editor of *JAMA,* did not agree, however. He was asked by Congressman Weiss at a subcommittee hearing, "Do you think that a journal should be willing to review two manuscripts with different results of the same study to determine which is correct?"

Lundberg replied: "It almost never happens, but if it were to happen, our answer is yes, the journal should receive whatever manuscripts it receives in good faith, proceed with the review process, and in the interest of the public, publish the most truthful, important articles possible without being particularly concerned as to what the institutional politics might be. Although downstream the institution might get kind of upset about that approach with the editor.

"I think the public interest comes first there, and the institution's interest is way down the line."

Responding to Cantekin

On January 25, 1988, Eugene Myers, MD, the chairman of Cantekin's department, sent a letter to the dean charging Cantekin with "research misconduct." The dean impaneled an ad hoc faculty review panel, chaired by a Pittsburgh professor of surgery, Charles G. Watson, MD, to investigate the charges. A transcript of the February 16, 1988, meeting of this committee raises questions about the panel's objectivity, however. One member asked Bluestone why the chairman of the department could not "deal with" Cantekin, and why Cantekin could not simply be put "in a closet." The transcript concludes with Dr. Bluestone saying that Cantekin is a "sick person" and he, Bluestone, as a medical doctor could not be "mad at" someone who is sick, but that he did feel "terrible" for Cantekin.[65]

On March 21, 1988, the Watson committee unanimously concluded that Cantekin appeared "guilty of several instances" of seriously breaching "academic integrity."[66]

The Watson committee claimed to find "serious breaches of research integrity" by Cantekin. The one listed first concerned the submission of his paper to the *NEJM*. The report says that Cantekin submitted his manuscript to the *NEJM* "almost simultaneously" with "Mandel et al." The committee stated that the *NEJM* editor wrote to the University and determined that "Mandel et al." was "*bona fide*" but that Cantekin's paper was not "institutionally certified." The committee restated Editor Relman's position concerning the importance of the principal investigator in deciding "how and when" to publish. The committee characterized Cantekin's actions as an "inappropriate expropriation of data" that indicated a "serious breach of research integrity."

Of course, Cantekin's cover letter to the *NEJM* made clear that he was not submitting his report as "the primary report." On the contrary, he was submitting his report, as a co-investigator, because the primary report in his judgment was false.

The University of Pittsburgh's Senate Tenure and Academic Freedom Committee (TAFC) also looked into the Watson committee report, and was dismayed by the Watson committee findings, commenting on their unbalanced character. The TAFC noted in a May 23, 1988, letter to Cantekin that "many" of the things known with certainty in the case "seem to have been ignored." Most of the members of the TAFC weren't medical research scientists, but they based their findings on the same examination of the documents detailed in this chapter. The TAFC found that the scientific community did not put the search for truth ahead of the search for alleged title to property or alleged, undocumentable "tradition."

Additionally, the TAFC wrote in its May 23 letter to Dr. Cantekin that it was concerned that in response to Cantekin's complaint, the Medical School had made "counter-charges" against him. The TAFC noted that this sort of action is regrettable because of its "chilling effect on open communications."

Cantekin had submitted his report to *JAMA* for publication, and initially they were interested in publishing the paper. But when Dr. Cantekin forwarded a copy of the Watson committee report to *JAMA*, the editor changed his mind about publication. In a letter of May 13, 1988, to Cantekin, the editor added the proviso that he would be "willing to reconsider" after the completion of the investigations by the NIH and the University.

As Congressman Weiss would later state, "[The University has] achieved what they want to do. They've shut up Dr. Cantekin."[67]

Retribution

Congressman Weiss's congressional report detailed Cantekin's subsequent harassment, the most serious of which were attempts to revoke his tenure and fire him from the University. The report states: "[O]n March 26, 1990, Dr. Cantekin received notification from Children's Hospital that 'Because of the renovation plans of the Hospital, to accommodate new and expanding services, it has become necessary to relocate your office' from the hospital to a windowless room above a supermarket. Dr. Cantekin protested this move, and on March 27, 1990, the General Counsel of the University assured Cantekin's lawyer that the move would not take place. However, the following morning, security personnel broke into Dr. Cantekin's office, packed his belongings, and moved them to the Giant Eagle supermarket."[68]

In his 1988 statement to the Weiss subcommittee, Dr. Bluestone stated that "no punitive action" had been taken against Cantekin as a consequence of his differing views on the studies or the accusations he had made. Bluestone insists that Cantekin's office was moved, along with those of four physicians, because of "critical space problems" at Children's hospital.

Cantekin testified that key research data was erased from his hard disk on Dr. Bluestone's orders. The Weiss Report elaborated: "Dr. Bluestone and his colleague claimed that Dr. Cantekin's data were merely moved from one data tape to another, that they were still available to him upon request, and that the moving of the data was not the decision of Dr. Bluestone. However, an engineer working on the project notified the subcommittee that he removed the data from Cantekin's hard disk on the direct orders of Dr. Bluestone. The data were not made available to Dr. Cantekin for more than two years. . . ."[69] When Cantekin finally did get access to his research data, he could only use it in connection with a hearing board proceeding because the Children's Hospital claimed it owned Cantekin's research.

After the Watson committee made its report on Dr. Cantekin's actions, the dean appointed a five-member hearing board, headed by Brian V. Jegasothy, MD. (Cantekin's lawyer objected to the members of the panel because of their affiliations to the University of Pittsburgh School of Medicine.)

The hearing board concluded that instead of writing a letter to the editor or a dissenting review, Cantekin tried to "derail the publications of his colleagues," writing not a dissent but a "parallel publication, fraudulently presenting the data as if he was responsible for it." According to the Hearing Board, Cantekin's actions amounted to "serious violations of research ethics" as well as "unethical practice, academic misconduct and breach of research integrity." He was also "uncollegial and unprofessional." The Board said he resorted to these activities when he could not accomplish what he wanted by "rational discourse."[70]

The dean "adopted" the board's decision and the senior vice president for Health Sciences "affirmed" it. Cantekin then appealed to University of Pittsburgh President Wesley W. Posvar, who appointed a five-person appeal panel chaired by George A.

Jeffrey, PhD, Professor Emeritus of Crystallography. Cantekin and his lawyer again strongly protested about one of the members, who had a joint appointment with the Medical School.

The appeal panel based its conclusion on "the undisputed facts" (which had already been disputed both by the University Faculty Senate's Tenure and Academic Freedom Committee (TAFC) and by the Weiss subcommittee). Some of the "undisputed facts" were even disputed by one member of the panel, Max A. Lauffer, PhD, who dissented. The remaining four panelists concluded that Dr. Cantekin was guilty of "serious violations of scientific and academic integrity."[71]

The appeal panel itself noted several "confounding aspects" in the matter, which the panel literally raised in the form of questions in its own report. If Cantekin was trying to get priority over "Mandel et al.," one would expect he would send it to another journal besides the *NEJM,* which was already considering "Mandel et al." Why did he not do so? If Cantekin was trying to grab credit from "Mandel et al.," why did Cantekin's cover letter specifically talk about "Mandel et al."?

The Jeffrey report then concluded: "We cannot explain Dr. Cantekin's motives or methods." Cantekin's lawyer, Robert Potter, wrote Pittsburgh University President Posvar: "The Committee's frank acknowledgment (which appears to be a form of postconviction hand washing) that it cannot square its conclusions with the incontrovertible facts is an acknowledgment that its conclusions are erroneous and cannot be defended as an impartial evaluation of the facts."[72]

In addition, the Appeal Panel also wrote that Cantekin had committed a serious academic breach by submitting an abstract concerning the AB-OME-I study "without recognition of or reference to" the authors of "Mandel et al."

Cantekin had explained what this was about in his *JAMA* article on peer review. "In September 1987, E.I.C. presented his dissenting views at the annual meeting of the American Academy of Otolaryngology, again citing the OMRC publication in the *NEJM* and comparing his analysis with the conclusions of the OMRC." At this meeting, Cantekin had also submitted an abstract of his talk. But Bluestone didn't want any citations of him, the center, or "Mandel et al." He wrote the Executive Vice President of the Academy, stating that he did "not give permission" for the published abstract to include any references to "this specific study" ["Mandel et al."], the Otitis Media Research Center, or the Department of Pediatric Otolaryngology of Children's Hospital.[73] So the Academy did not insert any of those references in the published abstracts although, as Cantekin's lawyer explained to President Posvar: "Dr. Cantekin's presentation was a point-by-point analysis of the differences and disagreements between "Mandel et al." and "Cantekin et al.," all of which was recorded by Dr. Myers who sat in the front row and took flash photographs of the presentation."[74]

Dr. Cantekin had written the abstract to conform to Dr. Bluestone's insistence that there be no written references to "Mandel et al." or the Center. Nonetheless, the Pittsburgh Medical School found Cantekin guilty for not making references to "Mandel et al." and the Center in his written abstract.

The President of the University of Pittsburgh wrote Cantekin following the report, informing him that he was adopting the Jeffrey report as his own decision, specifically excluding as part of his adoption, Dr. Max A. Lauffer's dissent. (Lauffer dissented for three reasons. First, Cantekin had submitted his paper to the *NEJM,* which was the same journal that was already considering the "*Mandel et al.*" manuscript. Cantekin's cover letter explained he was offering an "alternative interpretation." Second, Cantekin had offered coauthorships to Mandel and Bluestone. Both the offers and their rejections of the offers were in writing. Third, when Cantekin presented his paper at the Academy meeting, he had acknowledged "Mandel et al.," even if the abstract had not, for reasons we have seen).[75]

President Posvar remanded the matter to Dean Bernier "to consider sanctions." Dean Bernier recommended that Cantekin's tenure be revoked.[76]

However, revoking Cantekin's tenure required the approval of a committee that had almost the same composition as the Senate Tenure and Academic Freedom Committee. That committee, in a May 23, 1988 letter from Robert D. Mundell to Dr. Cantekin, had put in writing its exceedingly low opinion of the way the various Medical School's hearing boards had operated, describing their results as "less than unbiased." The TAFC wrote Cantekin offering further assistance if the Medical School's various proceedings continued as they had in the past.

Thomas Detre, MD, the senior vice president for Health Sciences wrote President Posvar on May 11, 1990, that there were not "definitive answers" to all of the issues Cantekin had brought up.

This justified a less onerous sanction than that of revoking Cantekin's tenure. Detre recommended five years probation. Detre also recommended that Cantekin be overseen for the five years by another committee appointed by the dean of the Medical School. The committee would be composed of faculty from the Medical School who specifically were not in Cantekin's department. What would this committee be on the lookout for? They would watch for acts of scientific misconduct or anything else that would "independently" be sufficient grounds for firing him from the university. In a May 18, 1990, letter to Cantekin, President Posvar wrote Cantekin that he had adopted Detre's recommendation.

Office of Scientific Integrity Report

Perhaps not too astonishingly, the OSI panel that neglected to investigate the anomalous placebo cure rate in AB-OME-I *did* investigate countercharges filed by Bluestone against Cantekin.

How the OSI handled what turned out to be an inquiry into Cantekin is worth looking at. The University of Pittsburgh had found Cantekin guilty for not referencing in the abstract presented to the Academy of Otolaryngology those things Bluestone had stated in writing he did not want referenced. The OSI report noted, "Thus, the panel believed there is evidence that the Cantekin et al. publication was perceived by some as the official publication of the AB-OME-I trial."[77] Cantekin wrote back to them that their "failure to have included [what the documents show] . . . in the Inquiry Report is further evidence of bias."

The OSI Report also stated "that the *NEJM* manuscript was misrepresented by the authors as the official report of AB-OME-I for publication." Cantekin responded, "OSI draws its information solely from documents provided by the University of Pittsburgh. . . ."

The OSI also suggested that the results in Cantekin's paper for various treatment groups should have been combined and subjected to statistical analysis and comparison. "This comparison was not made" wrote the OSI Report. Apparently the panel had not carefully read the paper. Cantekin wrote back to the NIH that "It is more than wrongheaded" for them to charge him with failing to do something when in fact he had done it and it was in the paper. He advised the panel to read Cantekin et al. until they got to table 4.

Two aspects of the panel's procedures are worth noting. First, Dr. Cantekin did not know he was even a subject of an inquiry until he showed up at an OSI panel meeting expecting to testify against Dr. Bluestone, only to discover that he himself (Cantekin) was the subject of inquiry. Second, the draft report not only was undated, but contained no signature. Cantekin had no idea who wrote it, until I interviewed the OSI's Dr. Suzanne W. Hadley, who said, "I wrote the report. . . . I got contributions I think from each member of the panel. I used some of them. Others I edited to ribbons. . . . But I put it together, and I take responsibility for that, and the OSI as a body does."

Based on this "careful" and "unbiased" analysis, the OSI report stated that Dr. Bluestone's activities "did not approach a charge of scientific misconduct." The report did, however, recommend that the OMRC be put under "special oversight" by the government funding agencies for five years. As of November 30, 1991, Dr. Bluestone continued to receive funding and had not been barred from further submissions to the funding agencies.[78] As for Dr. Cantekin, whom the university, in March 1990, had banished to a windowless room above a downtown supermarket and removed from all research, the Report "recommended" that he should be subjected to three years "oversight" by the University. The report did, however, also recommend that the university "move promptly" to find Cantekin a research position "so that he can return to his chosen career." The report "acknowledged" that Cantekin "identified legitimate concerns about OMRC research" but added that "his actions were at best uncollegial and at worst, they notably deviated from accepted scientific practices."

The report was then accepted by the NIH's acting director, William Raub. He wrote Dr. Cantekin on February 4, 1991, and stated, "I have accepted the OSI report and will ensure that the recommendations are implemented." Bernadine Healy, MD, the current head of the NIH, wrote Cantekin's lawyer a May 31, 1991, letter specifically referencing, enclosing a copy of, and presumably endorsing Dr. Raub's letter.

On December 18, 1991, *JAMA* finally published Cantekin's dissenting article, stating at the conclusion of an accompanying five-page editorial, "Looking back, surely publication of this dissenting view would have been very greatly to the benefit of all parties. We are now publishing it so that our readers can decide for themselves."

A CONCLUDING NOTE

The journal peer review process appears to systematically fail to correct error due to conflict of interest. Editors seem to defer to university claims of ownership and university investigatory procedures even if this has the effect of silencing the truth. Those who actually try to do anything about conflicts of interest and stand up for the integrity of scientific research, such as Dr. Cantekin, are opposed—initially by their own university, but ultimately by the very agency within the NIH that supposedly safeguards scientific integrity. Unfortunately, this does not bode well for science research or for the general public.

CHAPTER 7

THE PENTAGON SCIENCE GAME

The single biggest government expenditure in American science is the research and development (R&D) paid for by the Pentagon. Three underlying forces help to undermine scientific research for the military. The first of these involves *conflicts of interest,* particularly as seen on the Defense Science Board. The second is *secrecy.* Obviously, weapons development often has to be kept secret from foreign enemies, but Pentagon secrecy too often shields questionable practices from the congressional overseers who vote on Pentagon budgets. The third force distorting Pentagon research and development is the topsy-turvy practice of *concurrency.* Instead of research and development preceding production, R&D goes on simultaneously with production and sometimes even follows it. Concurrency is a major reason so many weapons fail to work properly when they are completed and why they are so vastly expensive.

THE EXCESSIVE PRICE OF RESEARCH AND DEVELOPMENT

"The big problem with Department of Defense R&D is out of control spending," said Thomas Amlie, a weapons developer who, until he recently retired, fought a losing battle to reduce spending by the U.S. Air Force. Before he became a Pentagon cost cutter, Amlie headed the China Lake weapons laboratory and was in charge of a team that developed a successful version of the Sidewinder, a heat seeking, air-to-air missile used by the U.S. Navy. Amlie, who has a PhD in electrical control systems from the University of Wisconsin, keeps a cigar-sized silver model of the Sidewinder on his desk. According to Amlie, the Sidewinder is an example of a successful weapon that deteriorated through repeated R&D. Although the first five or six models improved with each upgrade, eventually that pattern was reversed and the weapon became less effective with each new model. "The last successful upgrade was 'H,' about 1970 or 1971. 'L' and 'M' frequently don't go after the target, they track clouds. 'R' is really awful. It doubles costs and won't work at night because it operates on the visual spectrum," said Amlie.

The problems in R&D are the same problems that exist in defense purchasing. Both research and manufacturing are controlled by many of the same top Pentagon officials and are provided by the same Pentagon contractors. In a memo to senior Air Force officials, Amlie wrote: "This office has collected a great deal of data on why military equipment costs so much. . . . There are two basic reasons: Inefficiency and mark-up. We have examples of companies working at less than 10% labor efficiency and over 25,000% mark-up on labor."[1]

Pentagon spending on research and development is enormous when compared with the expenditures of other government agencies. The National Science Foundation (NSF) hands out roughly $1 billion (in constant 1982 dollars) each year to scientists and engineers. The National Institutes of Health (NIH) provides more than $6 billion for medical research. The Pentagon, by comparison, spent more than $40 billion in 1990 ($34 billion in constant 1982 dollars). To this must be added another $4.2 billion that the Department of Energy spent on R&D for nuclear weapons, military nuclear reactor research, and waste disposal.

Curiously enough, very little of this money is used for "basic research." The Pentagon spent a mere $724 million on basic research in 1990 (in 1982 dollars). The military spent another $1.8 billion on what Pentagon officials call "exploratory development" (in essence, applied research) and $1.4 billion for "advanced technology" development. Including the Strategic Defense Initiative,

"Star Wars," the Pentagon's total R&D spending came to about $7.2 billion dollars (more than the outlay of the NIH and the NSF). Most of the remaining $27.5 billion goes into the testing of specific items—what one analyst at the Congressional Research Service called "metal bending." The balance, more than $3.5 billion dollars, goes directly into the pockets of the major Pentagon contractors for "Independent Research and Development/Bid and Proposal" (IR&D/B&P). This is not a familiar term. A major report from Congress's Office of Technology Assessment (OTA) states, "There is no line item for IR&D funds in the defense budget."

Money for IR&D is, in fact, R&D funds tacked on as a fixed percentage of a specific contract with the Pentagon. Although the total amount of IR&D/B&P is small in comparison with other Pentagon funding for research and development, it is more than three times the amount given by Congress to the NSF. Viewed this way, the sum is huge. That such a large annual expenditure should have escaped public scrutiny is extraordinary. This chapter will focus on IR&D funding because it provides a perfect illustration of how politics, compromise, and fraud intrude into scientific research.

THE RESEARCH AND DEVELOPMENT CON GAME

Every six months the Defense Department's Inspector General publishes a list of criminal convictions or civil settlements under the heading "Significant Fraud Cases." During the period April 1 to September 30, 1989, for example, the list included Curtis Wright Flight Systems, Inc., which paid the government $1.4 million in a civil settlement concerning "inaccurate and incomplete cost and pricing data to the Air Force." Also on the list was Allied Bendix/Allied Signal, Guidance Systems Division, which settled with the Air Force, paying $1 million for "damages concerning falsification of test results for oil transmitters used in C-141 and B-52 aircraft engines." The Inspector General's list also included a host of smaller and lesser known companies and divisions. The same publication also listed major indictments, including Northrop and General Electric. In fact, at any given moment, roughly 60 of the top 100 Pentagon contractors are under criminal investigation by the Department of Defense. Yet these are the very corporations the Pentagon goes back to time and again when contracting for new projects.

In an attempt to defend themselves against charges of illegal spending, major Pentagon contractors set up their own self-policing organization in the mid-1980s. Called the Defense Industry Initiative on Business Ethics and Conduct, it put together a "self-disclosure" provision for companies that were willing to admit to illegalities in exchange for a lighter sentence. Of the top 100 Pentagon contractors, 54 refused to sign onto the program.

It is difficult to imagine, as an individual, doing repeat business with corporations that have been investigated, indicted, and convicted in major cases. But the 1989 Inspector General's Report lists an investigation of General Electric concerning an "inflated $244 million contract to deliver computer support systems to the Army." In February 1990, a jury convicted the company, and GE paid $16.1 million to the government and also agreed not to appeal. The same 1989 Inspector General's Report also listed a 167-count indictment against Northrop for "falsifying tests" and "conspiracy" concerning the Navy Harrier jet and the Air Force Air Launch Cruise Missile.[2] In February 1990, Northrop settled with the government for $17 million on the case.[3] Yet the Pentagon continues to give these corporations contracts and large infusions of cash—nominally for R&D (in the form of IR&D/B&P)—for which they are never held accountable.

IR&D is defined as a contractor's research and development costs that are not included in a contract or grant. The money is given to contractors automatically as a percentage of their billings for use in basic research, applied research, development, and systems and concept studies. It is a subsidy or entitlement given to contractors for research and development. According to most congressional investigators, about 2% to 3% of IR&D money is spent on basic research. A 1974 industry study gave a maximum estimate for basic research of 5%. The same study estimated that 30% of IR&D budgets go to applied research, 50% to development, and 15% for systems and formulation studies.[4] Roughly 90 companies receive 95% of IR&D funds, according to the Office of Technology Assessment. The remaining 5% is distributed to 13,000 firms. In effect, IR&D is a subsidy that excludes small companies.

Bid and Proposal cost, or B&P, is a separate funding category. It includes those costs incurred by contractors in preparing, submitting, and supporting bids and proposals (whether or not solicited) on potential government or nongovernment contracts. B&P includes no basic research. It is strictly the cost of studies to back up a bid and proposal. For example, B&P may include extensive wind tunnel analysis to show that an airplane wing will perform as a contractor says it will perform.

In practice, says a GAO investigator who specializes in this area (and who asked for anonymity), there is no objective way to distinguish between IR&D and B&P. A major report from the OTA says much the same thing. "In actual practice . . . companies do intermingle IR&D and B&P funds, and it is difficult for DOD [Department of Defense] to impose accountability and control mechanisms in this area."[5]

Cost Overruns

The problem with IR&D and B&P spending is that contractors often use the money to cover cost overruns. The DIVAD air defense system is an example of the use of an IR&D/B&P account to soak up cost overruns on other projects. The Justice Department took General Dynamics to court over the case but later dropped it when prosecutors decided that the Army had cooperated with General Dynamics in allowing the corporation to shift expenses on the weapon to its IR&D account.

Bruce Chafin, one of Congressman John Dingell's top investigators, wrote: "It is unclear whether anyone in the Army specifically gave the approval to these contractors to charge overhead accounts such as IR&D and B&P to cover the overrun. . . . However, it is clear from a number of post-indictment interviews that such charges were not unexpected."

Chafin's investigation disclosed other interesting features of Pentagon weapons development contracts and its link with IR&D: "It was revealed that there was common knowledge that the contract had been deliberately under funded and the contractor would far exceed the $39 million if they delivered a competitive prototype. The contract itself contained a 'best efforts' clause which would have excused the contractor from delivering any prototypes had they put forth their best efforts and spent the $39 million. However, the future potential of this contract was so great that most felt the contractors would spend 'their own money beyond the $39 million.'

"Numerous Army officials, in their post-indictment interviews, included IR&D and B&P as part of the 'contractors' own money'. . . . The most illustrative quote came from [a contract specialist at Rock Island Arsenal] who stated about the IR&D and B&P charges, 'Why not?'"[6]

This case is far from being the only one on record. According to the Deputy Inspector General, ". . . there are currently ongoing investigations concerning mischarging of costs by major defense contractors in connection with IR&D activities. These investigations involve allegations of mischarging of costs overruns from other contracts to IR&D, and IR&D costs exceeding limitations being charged to other contracts, and using historical costs which included mischarges not allowable as IR&D to establish current IR&D ceilings."[7]

Profit Advantages for Corporations

Most businesspeople would see the Pentagon's IR&D funding as a pretty sweet deal. Typically in private enterprise, research and development costs are accounted for as a general and administrative expense and are considered to be part of the cost of doing business. Rarely are such expenses charged to a specific project—general revenue covers this expense as part of overhead, along with other costs. Pentagon contractors too price their products based on the cost of producing the items, and add a substantial markup to cover overhead costs and provide for a profit. As Tom Amlie's supervisor, Ernest Fitzgerald, explained in an internal Air Force memorandum, "The major defense contractors that conduct the lion's share of IR&D make money primarily by selling allowable costs to the government." The IR&D and B&P money provided by the government is tacked onto the profit. "The IR&D allowances are grant-plus-percentage of grant arrangements," Fitzgerald explained in his memo to an Air Force general.[8]

The government not only pays for the contractors' R&D but also relinquishes all commercial rights to that R&D; any products that result from such research belong to the contractors. In addition, the contractors decide how to spend the money, which is why IR&D is called "independent" funding.

Even so, the Pentagon contractors complain about not getting everything they ask for in research and development expenses. Before the Pentagon authorizes any funds, the companies submit elaborate brochures to the government, explaining how they plan to spend the R&D money. The Pentagon then reviews the brochures and decides what percentage of the proposed work is relevant to Pentagon needs. Any R&D for which there is no specific contract must be paid for out of company profits. In 1989, the total IR&D and B&P requested by the top 131 Pentagon contractors came to more than $5.6 billion, but after looking over the glossy brochures, the Pentagon only paid out $3.6 billion.[9] Nonetheless, according to the OTA most of the defense contractors "receive almost automatic approval" of their IR&D plans.[10]

According to Amlie: "The size and cost of the IR&D reporting and proposals for contracts are a self-inflicted injury by the DOD. . . . They contain very little useful information and few people really read them except for the times when large groups of government employees are convened to evaluate proposals. I have been part of these groups on several occasions. The worst was in 1965 when over 100 Navy laboratory employees were ordered to Washington to evaluate proposals for [a specific weapon]. There were six proposals, several thousands of pages in all. In comparing the proposals with the [request for proposals], only one proposal could be called responsive. Trouble was, Navy brass had already decided that their favorite contractor would get the job in spite of the fact that his proposal was by far the least responsive and, as all the technical people agreed, totally incompetent."[11]

The RAND Report

What tangible fruit does the IR&D money provide? In 1986, at the direction of a House committee, the Under Secretary of Defense for Research and Engineering commissioned the RAND Corporation's National Defense Research Institute to investigate and report on the matter.

RAND reported its findings to the Senate Armed Services Committee on March 18, 1988. It was a one-sided hearing without so much as a single criticism of IR&D, though officials did discuss in passing whether the congressionally mandated cap on IR&D money should be lifted. (It was.)

"Congressional concern for justification [of IR&D] was very well placed," according to Dr. Michael Rich of RAND, who testified to Congress about earlier investigations into IR&D, "because past attempts to articulate why IR&D was important and why it was effective were really rather poor."[12] A 1975 General Accounting Office (GAO) report came to a similar conclusion when seeking the answer to the question, does IR&D money result in anything useful to the government? "It was not possible for us to make such a determination," said the GAO.[13] Rich, however, testified that the RAND report tracked down several examples of important new defense capabilities that were developed with IR&D. The only one he specifically cited to the Senate committee was the development of "the single crystal turbine blade" for jet engines, "developed by IR&D by both General Electric and Pratt & Whitney."[14]

Yet Amlie, one of America's foremost weapons designers, disagreed: "RAND claims that the single-crystal turbine blade (by Pratt & Whitney) was paid for by IR&D. A friend of mine . . . is a retired Air Force officer who worked as a jet engine designer at Pratt & Whitney after retirement. He said that everyone (GE, TRW, Pratt, etc.) was working on this because the metallurgy of the hot section, particularly the blades, was the only real problem left in turbine engine technology, and essentially defines the output of the engine. He said that everyone had to do it with or without IR&D. TRW claims credit for it."[15]

Fitzgerald noted the same thing: "The only hard example . . . RAND could give was their assertion that both General Electric and Pratt & Whitney had developed single-crystal turbine blades on IR&D. Our collective memories were that there had been many contributors to this technology for a very long time, and that in any event, *any* manufacturer of gas turbine engines should have been focusing on hot section metallurgy for the last 45 years, with or without IR&D."[16]

The RAND report claimed that the Pentagon's IR&D handout produces an in-house version of matching funds for *increases* in IR&D largess. "We found . . . that a dollar of *increased* IR&D cost recovery stimulates two dollars of IR&D effort [emphasis added]," according to RAND's Dr. Rich.[17]

Other observers disagree. A Congressional Research Service expert on IR&D said that the claims in RAND's report are based on budget figures put out by the contractors themselves and are not reliable.

The objectivity of the RAND report is questionable because this research division of RAND is literally "sponsored" by the Office of the Secretary of Defense (to the extent of $17.7 million in

1986).[18] In other words, the Research Institute, whose sole client is the Pentagon, was hired by the same Pentagon to write a report on whether or not the Pentagon was wasting its money. The government in an attempt to maintain honesty, provides oversight through the Under Secretary of Defense for Acquisition, with an advisory group that includes other senior DOD officials. The same person assigns the Deputy Under Secretary of Defense (Research and Advanced Technology) and the Deputy Assistant Secretary of Defense (Procurement) responsibilities for establishing IR&D policy. Therefore, the "oversight" of RAND's social science research into Pentagon spending comes from the very group that stands to be criticized, and as we have seen, RAND did not criticize the IR&D policies.

Amlie concluded that something like peer review would be a better way to award more than $3.5 billion in R&D money, arguing that the Pentagon should "let contracts based on the merits of the ideas rather than the size of the company." He added, "The DOD money for IR&D is viewed as a gift and is used to write proposals and warehouse idle engineers."[19]

THE TRUTH TELLERS

Fitzgerald and Amlie are thorns in the side of the Pentagon, and have played major roles in exposing defense waste. Ernest Fitzgerald, who is the nation's foremost expert on overspending in the Pentagon, has received high praise from those who try to stop such abuses. John Dingell said in a congressional hearing about Fitzgerald: "His analysis and his testimony for the subcommittee have led to significant changes in the way the Department of Energy operates, as well as termination of DOE's worst boondoggle, the Clinch River Breeder Reactor . . . His advice has turned out to be correct in each instance where he has advised the committee, and his record of being thorough and right is apparently at the root of his problem in terms of dealing with the Pentagon bureaucracy."[20] In contrast, Fitzgerald has been severely criticized by the big spenders. For example, Vern Orr, the Secretary of the Air Force during Reagan's weapons buildup, confirmed at a congressional hearing that he had once described Fitzgerald as "the most hated man in the Air Force."[21]

Years before, when Fitzgerald was an Air Force cost analyst, he had blown the whistle on the C-5A Transport. As a result, President Nixon fired him. Nixon was even recorded reminiscing on the White House tapes about it, in what may be the only known recording of an off-the-cuff presidential testimonial to a whistle-blower. "I said 'get rid of that son of a bitch.' You know, 'cause he is, he's been doin' this two or three times," Nixon said on the White House tapes.[22]

Fitzgerald fought a 10-year court battle to regain his job. His current job title is Management Systems Deputy, and he is, according to his own congressional testimony, "the highest person in the Air Force in the management controls field. My GS level is GS-18."[23] His current refusal to sign what he insists is an unconstitutional gag order signed by literally everyone else in the Pentagon and his belief that Pentagon money has to be regulated have heartened whistle-blowers throughout government.

Ironically, Fitzgerald and his immediate subordinate Amlie, whose jobs are to tell the truth about Air Force projects, work on the same floor as most of the Air Force's public relations officers, whose jobs, in many cases, are to present the Pentagon's views on the same projects. Uniformed public relations officers studiously avoid Fitzgerald in the fifth-floor corridors. Fitzgerald does not have altogether kind things to say about the public relations personnel either. Referring to the importance the Air Force places on public relations, Fitzgerald says, "A guy can *literally* become a two star general in bullshit around here."

HOW THE PENTAGON EVALUATES SCIENTIFIC PROJECTS

To evaluate scientific projects for military use, the Department of Defense uses two panels of "outside" experts. The more important panel is the Defense Science Board, which, as described in its own publicity materials "is the senior independent advisory body to the Department of Defense." The publicity materials state, "The [54] members-at-large are appointed for four year terms and are selected on the basis of their preeminence in the fields of science, engineering, technical, production and managerial skills . . . A group of senior consultants, also preeminent scientists and engineers, assist the Board in its deliberations." There are also a handful of ex officio members, including the heads of the corresponding science boards for the Navy, Army, Air Force, and Strategic Defense Initiative, and two other Pentagon organizations—the Defense Policy Board and the Defense Intelligence Agency Advisory Board.

Conflicts of Interest on the Defense Science Board

During much of the Reagan arms buildup, when a great deal of money was available to Pentagon contractors, the chairman of the Defense Science Board was the president of Martin Marietta, a major Pentagon contractor. He was ultimately succeeded, in 1989, by Dr. John Foster, described in 1983 by the Pentagon's Inspector General as "a brilliant scientist. . . . in the field of lasers especially. There is no question about Dr. Foster's capabilities as a scientist. . . ."

To this Congressman Jack Brooks of Texas replied, "Or his affiliation?"

Inspector General Sherrick had to agree, "Or his affiliation."

In 1983, Foster had been a vice president of TRW, another major Pentagon contractor. At the same time, Foster had chaired two Defense Science Board committees: the Committee on High Energy Lasers and the Committee on Space-based Laser Weapons. Although he retired from TRW in 1988, he was elected a director of that firm the same year. After his appointment, effective January 1, 1990, as chairman of the Defense Science Board, he continued as a consultant to TRW.[24]

Altogether, the list of members on the 1990 Defense Science Board included 29 who were affiliated with companies that do considerable business with the Pentagon, including a senior vice president of Rockwell International; the vice chairman and chief operating officer of Lockheed; the vice chairman of the Board of Science Applications International Corporation; the senior vice president and chief technical officer of Texas Instruments; the vice president of Science & Technology, United Technologies; the senior vice president of General Electric Aerospace; the executive vice president of Ford Aerospace; the chairman and chief executive officer of Hughes Aircraft; the senior vice president and general manager of IBM; and a senior vice president of ITT, who is also the president and chief executive officer of ITT Defense Technology Corporation.

"These individuals are corporate executives, not research scientists. . . . Should we ask corporate presidents for scientific advice?" asked Senator David Pryor about the members of the Defense Science Board.[25]

"The Defense Science Board is the *definition* of a conflict of interest," says Ernest Fitzgerald. Nine Republican congressmen from the minority side of the House Committee on Government Operations unanimously offered these conclusions in a report on the Defense Science Board:

"Although the Defense Science Board and the associated military advisory committees are made up of highly qualified experts, their findings have become overshadowed by the apparent conflicts-of-interest of some panel members. . . . even the indication of apparent conflicts-of-interest is disastrous to sustaining the credibility required from these groups in order to accept their findings on questions of billions of dollars in defense programs and in the direction of military research, development and procurement."[26]

Pentagon Computers

What brought the issue to a critical point was a Pentagon proposal, DOD Instruction 5000.5x, that dealt with a new computer policy establishing a formal system for the Pentagon to buy and make its own computer systems, even though the United States had, at the time, the world's most advanced computer industry. "If implemented, the policy will substantially benefit the large defense contractors and exclude the commercial computer industry from this multibillion dollar DOD computer market," said the House panel report. How did the Pentagon justify its policy? A special task force of the Defense Science Board recommended the move in a special report.

The General Accounting Office issued its own report asserting that the Defense Science Board's recommendation was ill-founded, and urging that it be dropped. The GAO suggested that the proposal would result in the Pentagon's buying dated technology, would restrict competition to those willing to install the less advanced technology, and would not allow the Pentagon to take advantage of the U.S. computer industry's future innovations.[27]

According to the House committee report, supported by both Democrats and Republicans, the DOD rejected the GAO's recommendation based on a report of a Defense Science Board's task force "composed mainly of defense industry representatives. . . . 7 of its 11 members had financial interests in one or more of the firms then holding contracts for pilot projects under the proposed new policy." Two members of the task force were from TRW. According to the House report, the man responsible for setting up the task force, the then Under Secretary of Defense for Research and Engineering, was a former executive vice president from TRW. His deputy, who was also the executive secretary to the task force, later became an employee of TRW.

Although the House Report does not indicate which other Pentagon contractors would benefit from the Defense Science Board's report, some task force members had affiliations with Bell Laboratories, Software Architecture and Engineering, Texas Instruments, Lockheed Missiles and Space Company, Control Data Corporation, Mitre Corporation, and the RAND Corporation.[28]

Surprisingly, Norman Augustine, who was then chairman of the Defense Science Board as well as president of Martin Marietta Aerospace, said, "[T]o date I have seen no evidence that there is a legal conflict of interest by the members of the task force."[29]

Paul Thayer, the Deputy Secretary of Defense and former president, chairman of the board and chief executive officer of LTV Corporation, one of the biggest Pentagon contractors, responded to the charge of conflict of interest with a circular argument, "[T]he advisory boards use outside experts who have commercial interests with organizations that do business with the Department of Defense. In most cases, their commercial experience is the source of their valuable expertise to the government."[30] In short, the appearance of a conflict of interest *is* their credential. "I may have the greatest conflict of interest of all," testified Thayer before the congressional committee.

"Yes you may," replied Jack Brooks.[31]

About four months later, Thayer resigned as deputy defense secretary, announcing that the Securities and Exchange Commission (SEC) intended to charge him with insider trading. The next year he pleaded guilty to obstructing justice in the case, and ultimately he agreed to pay the government more than $1 million to settle charges with the SEC. The next day, a federal judge sentenced him to a four-year jail term.

Regarding the task force's recommendation, Comptroller General Bowsher testified, "The procedures used by the task force in its deliberations did not assure adequate consideration of all points of view."[32]

An executive for Floating Point Systems Corporation protested that although there were many government witnesses in favor of 5000.5x, "[T]here were no Government witnesses allowed to testify or called to testify who had the opposite opinion."[33]

The Inspector General's report concluded that "the Board's compliance with governing policies, operating procedures, and legal requirements was clearly inadequate. Board task forces did not meet legal and procedural requirements, internal control procedures were not followed, statements of financial interest were not always obtained, minutes were lacking, proper notice of meetings was not given and the lack of documentation of the task force meetings left the objectivity of the Board recommendations open to question."[34]

Misrepresentation in Task Force Reports

The Inspector General found that conflict of interest was a pattern in other Pentagon task forces as well: "We looked in detail at six task forces and found that five of them did involve advice whose inherent nature could possibly result in advantageous situations for the firms with which their members were affiliated."[35]

The Inspector General does not have unlimited resources to beat down leads for every task force set up by the Defense Science Board, the number of which by far exceeded the 33 he initially tried to examine. In the end he had to be satisfied with a detailed examination of only 6. Between 1982 and 1988, the Defense Science Board issued 71 task force reports, according to documents they supplied. This is only a tiny fraction of the total number of Pentagon research reports over a five-year period, which according to the Pentagon's Inspector General came to 1,184 on laser weapons, 2,671 on cruise missiles, 11,535 on electronic warfare, and 12,493 on counter measures.[36]

One reason for such apparent conflicts of interest may be the method of choosing the task force members and the members of the Defense Science Board itself. The Pentagon's Inspector General noted, "Members were usually selected from the chairman's personal acquaintances or through the old-boy network within the military-industrial complex." A footnote to the report stated: "Of 64 personnel folders reviewed, 17 individuals had cited DSB members as references, and 37 individuals had been former government employees, 31 of which were former DOD employees. Furthermore, 19 of the former DOD employees came out of the Office of the Under Secretary of Defense for Research and Engineering."[37]

As a consequence of this investigation, the Pentagon promised to rectify the apparent conflicts of interest and bias at the Defense Science Board. To deal with conflicts of interest, they agreed to have potential task force members sign conflict of interest statements, which the Pentagon would review. To eliminate bias introduced by the old boy network, the Pentagon agreed to set up a list of possible task force members. And to remedy questionable behavior in general, the Pentagon provided the distinguished executives and scientists with their own ethics counselor.

The changes, however, seem to have been insufficient. When the Pentagon's Inspector General consulted the new master list of possible task force members, of the 199 listed, "121 still had direct ties to the DSB (42 were former members, associate members or senior consultants of the DSB, and 79 were proposed by current and former members, senior consultants and staff personnel); . . . 11 were recommended by senior officials from the Office of the Under Secretary of Defense for Research and Engineering; and . . . 32 individuals, all members of the Institute for Defense Analysis (IDA) [the RAND Corporation], were nominated by an IDA member who also nominated himself."[38]

The question of biased findings and political influence resurfaced four years later about Star Wars, the Strategic Defense Initiative. A task force report said that the then current deployment plan for the system known as "Brilliant Pebbles" was too "sketchy" to figure out how much the system would cost or what it would do. The report stated (as read aloud at a congressional hearing by Congressman James Spratt, Jr.): "As a consequence of current gaps in system design and key technologies. . . . [T]here is presently no way of confidently assessing (1) system performance against JCS [Joint Chiefs of Staff] requirements, (2) system cost, or (3) schedule."[39]

The task force also said that approval for deploying the system should be delayed "for the next year or two" until "gaps in system design and key [missile defense] technologies" could be filled. The items the report listed as "missing technology" were not mere details, but crucial elements of the project, such as the vulnerability of the space battle stations to attack from the ground, the ability to target "the rocket hard body in the presence of the rocket plume," and the infrared sensors' ability to "carry out discrimination against anything but the most primitive decoys." The task force also pointed out that the technology to manufacture the particular infrared sensors required "is not yet in hand."[40] Congressman Spratt read aloud the damning statement from the Defense Science Board to Robert L. Sproull, PhD, former president of the University of Rochester and a staunch advocate of SDI. Sproull was a member of SDIO's Scientific Advisory Panel, who protested, "Look, Mr. Chairman, you put me in a terrible position. I really don't know how I can disagree."

The task force chairman, Robert Everett, a former head of a big Pentagon contractor, MITRE Corporation, dropped those sections from the report before forwarding it on to the Undersecretary of Defense for Acquisition, which he had the legal right to do, even though he had not discussed the matter with everyone on the task force. Everett told a journalist, "Nobody put any pressure on me." He also said, however, that he had deleted the sections after discussing the report, but before delivering it, to Under Secretary of Defense Richard P. Godwin.[41] "Any changes made between previous versions and the . . . draft made available to Mr. Godwin are assumed to be internal deliberations of the Defense Science Board panel," said a Pentagon statement.[42]

Tony Battista, then of the House Armed Services Committee's Research and Development Subcommittee, and one of the most respected and powerful congressional staffers on Capital Hill, disagreed. When the alteration in the report had been revealed, he said: "What Mr. Spratt read to you is the Everett Panel Report to be considered by the Defense Acquisition Board. The conclusions, the recommendations that he came up with were deleted from the package that is currently being considered in their meetings. . . . That is a lack of candor."[43]

SECRET SCIENCE: HOW THE PENTAGON HIDES ITS R&D

The Laser Crosslink Satellite

The science the Pentagon pays for not only suffers from lack of proper peer review and conflicts of interest but also from its very secrecy. Projects can slip through the relevant congressional committees and be hidden in the vastness of the Pentagon budget. A little noted, but to those familiar with it, notorious example is McDonnell Douglas's Laser Crosslink Satellite to Satellite Communications system.

The McDonnell Douglas Laser Crosslink Satellite was designed to allow satellites orbiting the Earth to spot the missile plumes from nuclear attack and to communicate with each other by laser, in case enemy action destroys ground-based relay stations. Congress has spent more than $400 million on the project, most of it under production rather than R&D contracts. As of 1989, the Air Force had orbited one satellite specifically for a laser crosslink although the satellite did not have a laser crosslink in it because McDonnell Douglas was unable to make the system work.

While Congress assumed work on the failed project had ceased, McDonnell Douglas scientists and engineers secretly continued research and experimentation on the crosslink. "We thought the R&D project had been killed, because the thing doesn't work, but suddenly found it had somehow been slipped back in under a production budget line," said a congressional researcher.[44] The astonishing proof of this is contained in a report by Congressman John Dingell's Oversight and Investigations Subcommittee stating that the subcommittee "began its inquiry after learning that, contrary to plans, the Air Force orbited its . . . satellite Number 14 in June 1989 without the [laser crosslink] and also plans to launch its next three . . . satellites without the [laser crosslink].[45]

The Navy's A-12 Stealth Plane

Other programs have been "blacked out" from most members of Congress. These "Black Programs" are known only to key committee chairpersons. At least one Black Program was so disastrous that the Navy itself took action to bring the problem to light. On April 26, 1990, the Secretary of Defense publicly announced that he was continuing work on the A-12, the Navy's "Stealth" carrier-based attack plane. As a subsequent Navy inquiry stated, the Secretary "indicated that the A-12 would likely fly in early 1991, and he did not identify any impediments to completion of the FSD [full-scale development] effort within the scope of the current contract."[46]

On June 1, 1990, however, the contractors, General Dynamics and McDonnell Douglas, surprised the Secretary and "advised the Navy of a significant additional slip in the schedule for first flight, that the FSD would overrun the contract ceiling by an amount which the contractor team could not absorb, and that certain performance specifications of the contract could not be met."[47]

The Navy discovered that no planes were even remotely ready for test flight. Although the first test flight was at least 18 months away, the project was already $1 billion over the contracted budget. The plane itself had severe problems in at least three areas—antenna, engine, and weight. It was already *four tons* over the specification, "with resultant adverse impact on operational performance and the current A-12 structural design."

As early as April 25, 1990, one day before the Secretary of Defense had approved continued funding for the project, the contractors revealed to key Navy officials in a slide show that "weight and related performance (were) unachievable" in the current A-12 model. The Navy's official report explained that a slide "identified 'weight growth' as a 'major technical spec deficiency,' with impacts on single engine rate of climb, launch/arrest wind over deck, structure, future capabilities, and recurring/life cycle cost. While 'pursuing technical solutions to mitigate impact,' the slide indicated the 'ultimate' possibility of 'spec write-down-after flight test.'"[48]

The Navy's November 28, 1990, report revealed the elements leading to the A-12's downfall: "[T]he [contractor] team continually made best case projections of cost at completion based upon overly optimistic recovery plans and schedule assumptions." The Navy report also gave the exact reason for the overoptimism: "The evidence indicated that the contractor team perceived significant pressure from upper management throughout the performance of the FSD effort to maximize cash flow. Such pressure would create an incentive to be optimistic, inasmuch as progress payments would be subject to reduction in the event of a contractor or Government estimate of an overrun."[49]

The contractors were also "overly optimistic" about their manufacturing abilities, underestimating the difficulties in stabilizing the aircraft design and the impact that this would have on tooling and fabrication. "At best, these failures resulted from a plain lack of objectivity at the contractor team level, and wholly inadequate oversight by General Dynamics and McDonnell Douglas corporate management."[50] The Navy's program manager had accepted these unrealistic reports despite receiving less optimistic reports from cost analysts.

The program manager's unlikely expectations were not the only causes of difficulty. Secrecy also played a part. "[I]t appears that the special access nature of the A-12 Program was allowed to interfere with normal mechanisms for higher-level oversight of contractor cost performance."[51] In addition, the information for constructing the A-12 was stored away from highly secured areas, in a private facility on a mainframe computer. "In the absence of formal reporting, information was transmitted verbally only, in couched terms, and with no feedback to confirm that the data had been received or its implication understood."[52]

On January 7, 1991, the Secretary of Defense canceled the entire project, the most expensive cancellation in Pentagon history.

Black Programs and Courtesy Bids

There appears to be a great deal of deceit in the classification of Black Programs. According to Tony Battista, no more than 20% of such projects had genuine national security implications. He remembered one "white project"—not top secret—that had been killed while he was on the House Armed Services Committee Staff because it didn't work. "The next year it somehow got slipped back in as a Black Project under another name. We killed it again." When asked if this project, which he couldn't identify because it is still classified, would remain "dead" to further research and development under yet another name, Battista responded, "I don't know."

Battista is now a management consultant, a "Beltway Bandit" in the Washington idiom, who advises Pentagon contractors on different methods of Pentagon purchasing. Essentially, Battista advises the companies on how to sell effectively to the Pentagon.

The problems of the Black Programs, according to Battista and others, are not only that these programs circumvent the usual budgeting processes but also that they usually involve either sole source contractors or a limited number of bidders who can fix the bidding among themselves in a process called "courtesy" bidding, so that each contractor takes a turn as the low bidder. Here is a portion of a 1984 transcript of a secretly recorded conversation between a government informant wearing a wire and a man suspected of "courtesy bidding":

SUSPECT: But I'm not going to let you have it for 55, I'm gonna at least let you bid it up to 65 or 70 because I know I'm budgeted for 100 . . .

INFORMANT: Uh-hmm.

SUSPECT: And I've got other competition that's coming in at, at 75, 80, 82.

INFORMANT: So everybody looks good. We make money.

SUSPECT: Yeah.

In another section of the transcript, the suspect explained how he priced his products:

SUSPECT: . . . nobody questions dollars or anything like that. As long as I can show competition, whether it's true competition or courtesy competition or bullshit competition, you know. One of the things that we're getting right now is we don't have any in-house estimates, and we won't have for a year and a half, two years until we have made some of this stuff. So when they come down and they say hey we need a tool like this made, I say okay what do you guys think it would take to make it? Well we think maybe 200 hours of machining. Okay. What's our in-house rate? Seventy-seven dollars an hour. Okay multiply 200 times 77 to give me a not to exceed figure. . . . So if it comes out to a hundred thousand dollars if I can get that tool made for eighty thousand dollars I'm the best guy. I just saved twenty grand. Okay, by the same token you can look at that job and say I can make that thing for fifty-five thousand.[53]

With courtesy bidding practices, there is no incentive for companies to learn how to design and manufacture parts more efficiently or less expensively. According to Battista, however, even if companies do make progress in process research and development, "[a] lot of technological advances are not shared by the scientific community."

CONCURRENCY

One reason R&D fraud seems rare in Pentagon projects is that they seldom remain pure R&D projects. "Contractors try to move projects into production as soon as they possibly can," says a GAO expert. Starting up the production line and turning out the items before the product has been fully researched and developed is called "concurrent production." The official definition of concurrency, described in a report to the Under Secretary of Defense (Acquisition) is the "degree of *overlap* between the development and production process."[54] For some major weapons, however, including the B1-B bomber, the contractors had finished production and delivered the items but were still working on the research and development.

By restricting the definition of concurrency to the early production stages and pretending that the later stages, or even the postproduction period, do not involve significant R&D, the Pentagon and its contractors can easily misrepresent a project's riskiness and thus its level of likely cost overruns. The Under Secretary's report of April 10, 1990, stated that the Pentagon had 123 acquisition programs that qualified as "major projects," 20 of which had not yet gone into "full-scale development," and thus did not qualify for the definition of concurrency. Another 69 projects that were "past the low rate initial production phase" were, supposedly, too far along to qualify for the definition of concurrency. Nevertheless, some of these projects were still, in fact, undergoing major R&D even though they were not described in the report as being in concurrent development. The report thus could understate the true figure by claiming that only 34 projects were in concurrent development.[55] Whether concurrency is defined as an overlap of R&D and production, or as R&D during and *after* production, it is evident that concurrency is used on a staggering number of weapons systems, accounting in part for the high rate of inoperable machines as well as rising costs.

Determining the Effects of Concurrency

The damaging consequences of concurrency can only be *proved* on a case-by-case basis. The ultimate example is the B1-B bomber, which has been fully produced and is supposedly America's major bomber, although, it was not used in the Gulf War. "[A]ll of the aircraft are considered deficient," wrote Congressman John Conyers and Ranking Republican Frank Horton of the Legislation and National Security Subcommittee of the Committee on Government Operations. The main problem was "the lack of an operable Tail Warning Function (which is expected to provide detection of a missile behind the aircraft and to activate dispersal of chaff and flares)." Their May 6, 1991, letter to Les Aspin and William L. Dickinson listed some of the consequences of concurrent development in this case: "The Fully Mission Capable Rate (that rate which requires that all systems be operable) is currently zero. . . . The military specifications of the defensive avionics system have been reduced by 80 percent, yet it is uncertain if the system can achieve the new lower requirements. . . . Between October 1988 and December 1990, five B1-B bombers (two in flight) experienced a new problem: engine failures as a result of broken engine blades. . . . The anti-icing system for keeping the engine inlets free from ice build-up does not work and causes considerable maintenance problems. . . . Due to the possibility of induction icing in the engines, the B1-B is currently prevented from even running the engines on the ground over standing water, slush, or melting snow, between 20 and 47 degrees Fahrenheit."

The letter stated that estimates for fixing the engine blades run as high as $500 million, and fixing the anti-ice problem could cost another $200 million. The anti-ice mechanisms could not be installed before the year 2000. Fixing the defensive avionics system, "which would jam enemy radars," could cost an additional $500 million, if it were actually done. "The Air Force plan to solve the defensive avionics problem, the CORE program, has been terminated with no alternative solution in sight."

The Profit Motive

Some concurrency occurs because of technical problems that do not become apparent until after a project has begun, but on many Pentagon projects the problems go far beyond mere technical glitches—they actually are major conceptual and design flaws. A GAO report states: "[T]he practice increases the risk that systems will be produced with major flaws."[56]

The excuse for concurrency is the shorter development time for weapons. "The justification was usually that there was an imminent threat and the system was urgently required to counter the threat," said Charles A. Bowsher, the U.S. Comptroller General.[57] The ultimate motivation for concurrency, however, has more to do with the profits contractors receive when the weapons are produced. "Remember my theorem," Tony Battista wrote Congressman Ron Dellums, "for every dollar you spend today in R&D you generate a mortgage of $10 to $20 within five years." In other words, a contractor can get 20 times as much allowable costs from production as from R&D. A project that is thoroughly tested during its development might reveal that it is not feasible, or has significant problems, in which case production might cease or never even begin, and thus the project would offer no large increase in allowable costs.

Vice admiral Robert R. Monroe explained the motivation of the contractors in rushing into production: "Those who favor programs urge more concurrency, because most of the serious problems in a system are identified in the T&E (testing and evaluation) periods that occur late in development, and if the system is already in production at the time these come to light, the program will have gained enough constituencies and enough momentum to survive. Those who *oppose* the program urge less concurrency, believing that this will give them additional opportunities to slow or kill the system before the startup of a production line makes this virtually impossible. . . . What we are really talking about in concurrency discussions is how much T&E data will be available at the time production decisions are made."[58]

Problems with Operational Testing

Testing done once the project is in production is not usually called developmental testing (testing to see if the unit performs up to specifications). Testing of a completed prototype, before production, and testing of a weapon, during or after the beginning of production, is called *operational* testing (testing to evaluate whether the weapon functions effectively, is maintainable, can be used by typical personnel, and works against an enemy trying to foil it).[59]

Developmental and operational testing may be combined, according to the GAO in a 1985 report, "when significant cost and time benefits will result, provided that the necessary resources, test conditions, and test data can be obtained."[60] If the weapon involves a high degree of concurrency, however, it may never go through adequate developmental testing to allow for operational testing. According to the same 1985 GAO report, the planes used in the operational flight tests of the B1-B were fitted with an avionics countermeasures system that was not even sufficiently developed for operational testing.[61]

There is often not enough information, to administer all the proper tests. For example, the GAO noted with regard to putting ICBMs on rail cars, "No OT&E [operational testing and evaluation] of the complete weapon system (missiles and rail cars) will have been conducted prior to the initial production decision."[62] The GAO listed seven other programs for which low-rate initial production had been authorized without operational testing and evaluation: the ALQ-165 Airborne Self-Protection Jammer; the SQS-53C Sonar; the E-6A Aircraft; the MK-45 Capsule Launching System; the Ocean Surveillance Information Baseline Upgrade; the TB-23 Accelerated Thinline Towed Array; the AN/BSY-2 Submarine Combat System.[63]

The absence of operational testing, or utterly inadequate testing, is also a reason for waste in major, hugely expensive weapons. "The B1-B did not begin operational test and evaluation until 3 years after the October 1981 production decision," reported Frank Conahan, Assistant Comptroller General to Senator Nunn's Senate Armed Services Committee. "Despite costly attempts to fix its critical avionics subsystem, the plane will not do what it was expected to do."[64] The SSN-21 Seawolf attack submarines had similar problems of a rushed schedule without adequate testing. "As many as 15 of 29 planned SSN-21's, worth more than $21 billion, are to be on contract or under construction before the first ship is available for operational testing," said Conahan. A major element of the Seawolf that has not yet been tested, and will not be tested until two years after it is delivered, is the AN/BSY-2, "one of the most technically challenging and complex software developments for a submarine," according to Conahan. The software involves 3.2 million lines of computer code. If it fails to work, the submarines may have to be redesigned. Conahan warned that this "could further delay deliveries and increase costs. And this has already occurred."

The B-2 has been the subject of a number of GAO reports. This plane has not been put through operational tests on its main claim, invisibility—most likely because it would fail the tests. The GAO's Frank Conahan testified, "Under the DOD acquisition plan . . .

over $48 billion would be appropriated before anyone knows whether this airplane will do its job. . . . We remain concerned . . . that production of this plane is continuing without adequate assurance that it can perform its mission."[65]

Tony Battista spelled out the problem in a July 24, 1989, letter to Congressman Ron Dellums, written at the congressman's request and widely circulated in Washington: "[T]he B-2 does not in my opinion render Soviet air defense systems obsolete. Without getting into Soviet technologies and systems there are RF technologies that have been around for over three decades that can detect objects with a cross section advertised by the Air Force with more than adequate reaction time. . . . let me suggest to you that you require the Air Force to fly an augmented B-2 against Soviet radar systems that we have or can acquire or even against some FAA radars such as the Air Route Surveillance Radar (ARSR) mod 2 or 3."

Dealing with the Consequences of Concurrency

What happens to all the defective equipment? According to Vice Admiral Monroe: "[T]he first few years of production runs will either have to be scrapped or reworked and retrofitted at great expense to the government (or—worse—they will be deployed with major performance shortcomings, vulnerabilities, and serious reliability/maintainability/availability problems)."[66]

The scenario that Monroe envisioned regarding the distribution of inoperable or deficient equipment to front-line troops is not uncommon. The same problem occurred with the Apache helicopter in the Gulf after the invasion of Kuwait. No one should have been surprised, however, for in 1981, the Army Logistics Evaluation Agency, according to the GAO, had "reported reliability problems with components, among them the targeting sensors and the 30-mm gun." Based on those evaluations as well as others, the GAO recommended that the Pentagon delay production of the Apache until key problems were resolved. A number of the problems were not fixed, and the Pentagon began full-speed production. In 1990 the GAO issued a report titled, "Apache Helicopter—Serious Logistical Support Problems Must Be Solved to Realize Combat Potential." By this time the Pentagon had acquired more than 600 Apache helicopters, with the intent of buying 807 in all. The machine had a number of flaws. The main rotor blade, for example, which was expected to last 1500 hours, failed after 164 hours. The 30-mm gun, identified as having problems in 1981, supposedly could fire a minimum of 3,838 rounds before it jammed or a component broke; the gun was able to fire only 1,048 rounds before jamming or breaking.[67] "Army maintenance personnel can usually repair the gun by replacing failed components with new ones," said the GAO.

The reason fraudulent developmental testing is so rarely reported may be because there is so little developmental testing on which to report. There is documented evidence for this suspicion in a 1974 GAO report that detailed development tests on the Navy's antisubmarine devices. The GAO, after discovering that the developmental testing was inadequate in every stage of the process except for the amount of money allotted for testing, stated, "The development projects we reviewed needed improvements in planning and performing development testing and reporting and using development test results."

An example of insufficient developmental testing is apparent in the new Sonobuoy, a device that, when dropped into the sea by antisubmarine airplanes, sends out sonar signals that are picked up by the plane. The Sonobuoy tested was the AN/SSQ-62, a successor to the earlier AN/SSQ-50. Scientists testing the new version skipped testing a wide range of this device's supposed capabilities "on the premise that other operational requirements . . . had been proven feasible by the . . . 50 sonobuoy's performance. . . ." The earlier AN/SSQ-50 had not been fully tested, however, and was only 69% reliable at best—far below operational requirements. In addition, the GAO noted that the 62 design is different from the 50 design and "the [subsequent] Operational Test and Evaluation Force was not satisfied with the . . . 50 system's performance capabilities for several of the . . . 62's characteristics which were not tested."[68] The GAO concluded that necessary tests were not always done "or their planned coverage was substantially reduced." As a result, "problems were often not resolved and continued into operational testing of the equipment."[69]

Questionable practices in operational testing is a subject of considerable controversy. In the testing of the MX missile, Frank W. Lynch, vice chairman of the Northrop Corporation, which manufactures the MX, affirmed that concurrent production was common: "We are building both full scale engineering development and 'production models.' We are evolving the process of building these all together and at the same time."[70]

These ICBMs were supposed to use nuclear warheads to destroy "hardened" Soviet missiles (missiles in their concrete silos). This mission requires that the missiles are accurate in reaching their targets. Lynch explained: "[W]hen [the Air Force] put a number down, that is a circle within which 1/2 or 50 percent (of the warheads) will fall, and that is the definition of accuracy. There is a second circle which is an outer circle that all must fall within.[71]

This definition of accuracy allows for a very large percentage of error. If half the thermonuclear warheads on the missiles hit inside the first circle, the test is considered 100% successful. "The half that blow up other places will be totally disregarded," said Dingell subcommittee staffer Bruce Chafin in an interview.

In the operational testing, according to Congressman Ron Wyden, the data make it clear that the accuracy of the MX missile is getting worse. "The Air Force has admitted that the accuracy of the weapon has experienced a significant decline." What happens to these badly guided missiles? They are being shipped to their silos and deployed. The lack of response to actual testing criteria has created potentially irreparable and catastrophic damage.

DEFICIENCIES IN PERFORMANCE OF TESTS

The actual tests are often performed in a careless fashion and under unrealistic conditions. In 1988, the GAO reported on operational tests of six major weapons systems, two each from the Navy, Army, and Air Force. The GAO's summary conclusions listed a number of deficiencies in the tests, including: "(1) the absence or significant underrepresentation of countermeasures (communication jamming, radar jamming, electro-optical countermeasures); (2) the use of tactics that facilitate performance during the test but are incompatible with system survivability in [combat] . . . ; (3) the use of targets that are hotter, slower, higher, more plentiful, less maneuverable, more likely to be stationary, or less likely to be camouflaged than DOD sources indicate would frequently be the case in combat, and (4) . . . the outer edges of the specified performance envelope were not tested. . . . [I]n one case, needed meteorological data were obtained by a telephone call to the target area, an implausible method of data collection in wartime. . . . As a result, estimates of performance . . . tend to be biased upward, and performance under more realistic stress conditions remains unknown."[72]

Not only are the test conditions unsuitable for collecting accurate data, the information is manipulated even as it is being recorded. The GAO identified 22 cases where test reports drew false conclusions. In a document titled "Weapon Performance," the GAO revealed that "the Army's test report on the Multiple Launch Rocket System did not adequately describe . . . a problem with the ammunition resupply trailer—namely, when carrying a full load of ammunition and traveling at normal speed it tipped over."[73]

SUMMARY

Conflicts of interest, secrecy, and concurrency have developed in the Pentagon not because of the need for national security, but because of greed. Until the Department of Defense resolves these problems, its R&D will continue to produce massively overpriced and often substandard equipment. The Pentagon officials who control defense purchases will continue the practice of repeatedly reducing the specifications of weapons and claiming that that is what they wanted after all. Ernest Fitzgerald summed up the basic result of this R&D process when he stated, "It's not that the weapon fits the specification, it's that the specification fits the weapon."

CHAPTER 8

WHERE DO WE GO FROM HERE?

This book has described a wide range of political manipulation, compromise, and outright fraud in a variety of scientific fields—early man research, earthquake-engineering research, weapons developments, new pharmaceuticals and medical devices, and basic medical research. The conflicts of interest, institutionally abetted or officially sanctioned, reveal a consistent but unfortunate picture of how the science world works in the United States.

THE UNIVERSITY AND THE PENTAGON

Much of the science discussed here has had a Pentagon connection. This has continued to be true in scandals disclosed in the 1990s regarding overhead billing and university research. In the spring of 1991, Congressman John Dingell held two congressional subcommittee hearings that directly linked top university officials and their universities with fiscal improprieties in scientific research.

The first hearing focused on Stanford University. Dingell's hearing revealed that of the roughly $240 million per year that the government pays Stanford for conducting research, 74 percent has been spent, not to support specific research projects, but to pay "indirect costs" of the university. A GAO auditor explained, "Indirect costs are those that cannot be identified with a particular project or activity. These would include such costs as utility expenses, depreciation of buildings, and general university administrative costs."[1]

Much of this outlay was not overhead on science research contracts at all. Dingell enumerated a few of the items: "[Former Defense] Secretary Weinberger look[s] rather poor with his $600 toilet seat. Stanford recently purchased an early 19th Century Italian fruitwood commode. That was $1200 and was subsidized by the taxpayers. The taxpayers also contributed to the enlarging of Dr. Kennedy's [Stanford president] bed; $7,000 for sheets for the enlarged bed, and the purchase of Voltaire chairs from Pierre Deux, at $1500 each and a pair of George II lead urns at a special price, $12,084. They also charged the taxpayers $400 for flowers which were used in connection with the dedication of the Stanford horse stables."[2]

Admiral William C. Miller, Chief of Naval Research, Office of Naval Research (ONR), explained how his agency's oversight procedure is supposed to work. The ONR provides government oversight on federal grants, from all federal agencies, for 40 universities. "Indirect costs or overhead rates at a particular university are set for a forthcoming fiscal year through negotiations with the onsite [ONR] administrative Contracting Officer and the University."[3] However, most of Stanford's questionable indirect costs were permitted by the Navy's contracting officer under "memorandums of understanding," or MOUs. These MOUs permit overhead allocations that otherwise would not be permitted by the regulations. At the time the Stanford scandal surfaced, the university had 90 of these memorandums actively in force. In contrast, Cal Tech had the next highest number of MOUs, with 13, and the Massachusetts Institute of Technology (MIT), was third, with 7.[4]

Admiral Miller added, "After the year is completed, the Government is supposed to come back and audit the actual expenditures of the University and to adjust the rate and balance the books, if you will, regarding any over or under-allocation and collection by the University." Auditing these contracts should have involved the Defense Contract Audit Agency (DCAA), according to Admiral Miller: "The Office of Naval Research should have requested that DCAA conduct the audits. We do not have audit staff. DCAA shares this responsibility for the Department of Defense with us." (Some of the contracts associated with major Pentagon scandals discussed in Chapter 7 were, in fact, "audited" without comment by the DCCA.)

The DCAA only performs its university audits if the Navy requests them to do so. The defense director of the DCAA, Fred Newton, testified, "We did some auditing during the 1980s but of the allocation procedures that are the subject of these memorandas of understanding, we were not asked to audit those."[5]

Admiral Miller confirmed, "[A]udits had not been conducted of the actual costs incurred by the University since 1981."

This nearly total lack of Pentagon controls coincided with the election of a President from California who immediately launched a massive arms buildup that resulted in major contracts to universities.

Asked by Congressman Rowland if he would "draw a parallel" between Stanford and one of the most notorious of the Pentagon contractors, General Dynamics, Newton replied, "I would draw a parallel between General Dynamics of the early 1980s and the other major contractors who were abusing the Government contracts, yes, sir."

Rowland asked, "Is Stanford worse than General Dynamics today?"

Newton replied, "There was more money at General Dynamics. So, I would have to say that is worse."[6]

Asked who he felt was responsible for what had been happening at Stanford, Admiral Miller replied, "Sir, from my command experience, I am held responsible for my command in the Navy and in that regard, I would believe that the President of Stanford University is responsible for the entire operations of that University as well."[7]

At the time of the hearing, the General Accounting Office (GAO) auditors had only been looking into the questionable indirect costs for a short time but had already found some big ticket items that should not have been there. The total amount of dubious billings was in the millions and growing. For example, Stanford, which had acquired a 73-foot yacht as a gift, advertised it in its Sailing Program brochure, rigged it with a jacuzzi, and depreciated it as overhead on scientific contracts. Stanford owned a shopping center as well, the administrative salaries and expenses of which the university charged as overhead on science contracts in the amount of $185,000. Stanford also had an unusual method of depreciation. According to the testimony of the GAO auditor on the case: "Stanford used an accelerated method of depreciation for buildings and improvements, rather than the . . . prescribed straight-line method, even though it has not provided adequate justification for doing so. . . . Stanford was able to recover $4.6 million more in depreciation charges in fiscal year 1986 than it would have been allowed under the . . . standard straight-line method."[8]

This phenomenal amount of overhead, which was slipped in under the label of "scientific research," began at the time of the Vietnam War and the Apollo moon-shot program. The GAO auditor explained: "Until the mid-1960s, Federal reimbursement for indirect costs was limited to 20 percent. In 1966, the limit was removed. By 1990, the average indirect cost rate charged by universities had risen to about 50 percent."[9] Each one percent increase was worth another million dollars to Stanford.

By Dingell's second hearing, in May 1991, the estimate for Stanford's overbilling on questionable overhead during the preceding 10 years was estimated to be between $160 million and $200 million. Although Stanford seemed to win first prize in the "dubious overhead" contest, other universities were major players, including Dartmouth, Duke, Cornell, and MIT. Dingell noted that at Dartmouth, the federal government paid out "some $20,490 for the chauffeuring of the President and his wife," and at MIT the taxpayers paid for "the dues of the President at the Cosmos Club . . . and the salary of the President's cook."

SELF-REGULATION?

Members of the scientific community are continually advocating changes in the system to answer some of the issues raised in the previous chapters. Invariably, these changes involve alterations or refinements in the current system of "self-regulation." A typical example concerns conflicts of interest in the clinical testing of new pharmaceuticals or medical devices.

Arnold Relman, MD, editor of *The New England Journal of Medicine,* in an April 6, 1989, editorial, warned that there is a sure way for the scientific community to get more cases of dubious behavior, which will result in more bad publicity. The bad scenario will result if individual scientists and corporations are completely left to their own "private discretion" as to what they should or should not do. Relman's proposals to avoid that situation began with full disclosure to both the journals and the funding institutions of "all financial ties" between scientists and the "products and procedures they are investigating." He then endorsed "self-regulation by investigators,"[10] specifically that detailed in the same issue of the *NEJM* in an article signed by, among others, Bernadine Healy, MD, the current head of the NIH.

The guidelines state: "Investigators involved in the Post-CABG study will not buy, sell or hold stock or stock options in any of the companies providing or distributing medication under study for the following periods: from the time the recruitment of patients for the trial begins until funding for the study in the investigator's unit ends and the results are made public; or from the time recruitment of patients for the trial begins until the investigator's active and personal involvement in the study or the involvement of the institution conducting the study (or both) ends.

"Certain other activities are not viewed as constituting conflicts of interest but must be reported annually to the coordinating center; the participation of investigators in educational activities supported by the companies; the participation of investigators in other research projects supported by the companies on issues not related to the products in the trial and for which there is no financial payment or other compensation. . . ."[11]

Relman then recommended "a broader and more institutionalized approach." He quotes, with apparent approval, a proposed NIH guideline that states that researchers it funds "will not have financial interests in organizations or entities that produce drugs, devices, or other inventions studied in a controlled clinical trial."[12]

These may well be laudable "guidelines," but they do not impose penalties if the guidelines are ignored, nor do they indicate who would enforce the sanctions.

The evidence seems to show that too often the members of the scientific community who strongly advocate self-regulation are among those who do the pushing and shoving when an honest scientist steps forward to expose fraud. Since science in the United States, and probably most other places, is organized hierarchically, self-regulation thus far has meant that the people in the top ranks of science do most of the regulating, with the results documented in this book.

BUREAUCRATIC REGULATION

Self-regulation is obviously an institutionally weaker system of regulation than the bureaucratic regulation imposed, for example, by the National Institutes of Health (NIH). Documentation exists on how bureaucratic regulation functions in an area where it might be expected to function at its best—the testing of potentially lethal vaccines on healthy, human subjects. If the regulation fails in that critical area, where would it be effective? A current, documented case of serious failure concerns NIH scientist Dr. Robert Gallo in his collaboration with French scientist Daniel Zagury.

On July 6, 1990, the NIH received a long fax from *Chicago Tribune* journalist John Crewdson, who had investigated Gallo's claim to have isolated the acquired immuno deficiency syndrome (AIDS) virus. Crewdson had followed up another story about Gallo that involved testing humans in Zaire and Paris with a possible AIDS vaccine developed by Zagury in collaboration with Gallo. When Crewdson could not get answers to some questions about Gallo from the National Cancer Institute (NCI), he sent his questions to the NIH press officer. Crewdson wanted to know why required documents were not on file at the NIH to authorize the use of NIH research materials in these human testings. He had been informed, he stated, that the government of Zaire had "made available" political prisoners in Katanga province for research by Dr. Zagury and those he worked with. Crewdson also said he had information concerning testing on humans. Specifically, alleged Crewdson, Zagury and others had not adhered to "generally recognized" informed consent procedures and posttrial counselling.[13]

The NIH investigated through its Office for Protection from Research Risks (OPRR) and then issued a confidential report on May 31, 1991, containing several interesting findings and recommendations. First, it noted that Department of Health and Human Services (HHS) regulations for tests involving human subjects applied whether the testing was done in the United States or elsewhere in the world. It found that most of the relevant experimental protocols and required filings had not been done. Furthermore, it found that the relevant NIH official, who was subsequently replaced, "had no direct authority over (and little opportunity to influence) intramural scientists who were not actually conducting research in the NIH Clinical Center."[14] So if the experiments were conducted in Paris, let alone Zaire, the NIH official could not do much.

The OPRR then required that all ongoing research involving NIH scientists with organizations linked to Zagury must receive reapproval. The order specifically referred to any "scientist affiliated with the Université Pierre et Marie Curie (France) or its associated laboratories, clinics, and hospitals (including, but not limited to, Assistance Publique Hôpitaux de Paris and Hôpital Saint Antoine)" as well as scientists affiliated with two research institutions in Zaire. Up to that point, Gallo's collaboration with Zagury had resulted in the publication of 14 papers "reporting research," according to the OPRR, "that appears to have involved human subjects."

The same month that Crewdson faxed the NIH, Gallo and Zagury published a report on their human experiments in the British medical journal, *The Lancet.* That report stated that none of the 19 human subjects had died; however, two died by the time the article appeared, and a third died later. They had been injected with vaccinia, the virus that causes cowpox. The NIH had genetically altered this virus by adding portions of the AIDS virus. According to a French government report, one of the three persons died as a direct consequence of receiving the altered vaccinia.[15]

The OPRR report described these deaths: "Dr. Picard [another French scientist] and Dr. Zigury have acknowledged to OPRR that three of the individuals enrolled in the trial for 'compassionate' reasons died following the experimental treatment, although the immediate cause of death may not be precisely known. All three individuals suffered injection-related subcutaneous necrosis that developed after they received local, as opposed to intravenous, injections. Two deaths involved subcutaneous injection, and one involved intramuscular injection. Dr. Picard has stated to OPRR that the experimental treatment shortened the lives of these three patients subsequent to the onset of the injection-related necrosis.

"Dr. Picard has indicated that she gave a formal presentation describing these deaths at the annual Laboratory Meeting organized by NIH scientist Dr. Robert Gallo in August 1990. Dr. Gallo, however, has stated that he first became aware of the deaths in January 1991, and at that time did not realize that the deaths involved research subjects. Dr. Picard and Dr. Zigury maintain that the deaths occurred after the submission of the 1990 publication."[16]

Perhaps the most hair-raising of the experiments was one that, according to the OPRR, "involved immunization of healthy, HIV-seronegative volunteers with synthetic HIV peptides." Supposedly, these "volunteers" were to sign a paper that clearly informed them of the horrible potential consequences. According to the OPRR report, however, "The latter information, intended to inform volunteers that they would become seropositive as a result of participation, does not appear in a (signed) consent document that has been obtained by OPRR." Were NIH scientists involved? The OPRR report states: "Although NIH scientists have denied direct involvement in this project, Dr. Daniel Zagury, the primary basic science investigator, has acknowledged to OPRR that reagents supplied by NIH scientists have been used for *in vitro* analysis of the blood samples. . . . Dr. Jean-Claude Imbert, the head of the hospital department in which this project was conducted, has publicly characterized this project as involving intellectual collaboration with NCI scientists."

The OPRR report had severe criticism for the NIH's system of oversight through its Institute Clinical Research Subpanel (ICRS), the review board that is supposed to review and, if it chooses, approve any activities involving human subjects: "NIH must be faulted for created an administrative structure (or vacuum) that has resulted in widespread confusion within NIH about the differing responsibilities of OPRR and the NIH ICRS system." What the report seems to suggest is that the American scientists figured that if the French did the actual injecting outside the United States, the Americans were off the hook. The report stated: "These scientists were unaware of the regulatory definition of human subjects and assumed that they had no responsibilities in this area as long as they did not directly inject human beings with experimental materials. Some seemed to believe that compliance with foreign standards was all that was required."[17]

The OPRR panel summarized its findings with regard to the NIH's bureaucratic jumble: "Generally, the OPRR investigation has revealed a disjointed, compartmentalized system of human subject protections within the NIH intramural research community. Lack of centralized and authoritative oversight of research activities covered by HHS human subjects regulations has resulted in uncertainty at all levels of the intramural community regarding individual and institutional responsibilities under the NIH Multiple Project Human Subjects Assurance."

In other words, the NIH has been incapable of applying its own regulations. Given this documented fact concerning bureaucratic enforcement, how much confidence can anyone place in such control? And would a *voluntary system of self-regulation,* which by definition lacks (1) oversight, (2) enforcement, (3) due process, (4) sanctions, and (5) remedies, be any more effective?

At least, however, the bureaucracy can keep close watch on one or two individual scientists. In a section titled "Final OPRR Action" the report states: "Until these scientists have established a record of strict compliance with HHS Regulations for the Protection of Human Research Subjects, (a) all human subject research activities involving Dr. Robert Gallo with investigators outside the NIH shall be forwarded to OPRR for additional review and prior approval; (b) no research activities (including shipment of research materials or instruction in research technology) involving Dr. Daniel Zagury shall be permitted by any HHS component or employee without the prior written approval of OPRR."[18]

The report recommended that special regulations be set up to monitor Gallo and Zagury's future research: "NIH should develop special administrative procedures to guarantee compliance with HHS and NIH human subjects requirements in any NIH-supported research activities in which either Dr. Robert Gallo or Dr. Daniel Zagury is involved."

Both Dr. Zagury and Dr. Picard later said that the report's questions concerning their work and the participation of the NIH researchers "are entirely without foundation." Zagury maintains that the work upheld the ethical and regulatory guidelines of both Zaire and France, and that the NIH did not necessarily have oversight jurisdiction over it. French regulators, he maintained, had oversight authority over his research.

CONGRESSIONAL OVERSIGHT

If self-regulation seems to mean little or no regulation, and bureaucratic oversight is ineffective, what about congressional oversight? Unfortunately, it, too, is not an effective, long-term solution. By the end of June 1991, John Dingell had held four hearings on the Baltimore case and was considering one or two more. As a result, the truth did come out and Margot O'Toole was vindicated. Such efforts, however, are few and far between. Ted Weiss, for example, held two superb hearings on the Bluestone case, but Dr. Erdem Cantekin remained an excommunicant in the University of Pittsburgh School of Medicine. In the Bruening case, Robert Sprague had his grant cut short, and he had to resume active research as a co-investigator on someone else's grant at another funding agency.

Congress can and does do a superb job in bringing to light a few cases of scientific fraud, but it does not have the resources to investigate many cases nor does it have the power to penalize the guilty or to provide a remedy for those wronged. In an interview, Dingell's top investigator, Peter Stockton, expressed a genuine wish to do more but did not think it would be possible with the subcommittee's resources—which are the most extensive in Congress. Stockton said: "Walter [Stewart] is always coming at me with this case or that. He always says 'It won't take much time,' but he knows damn well I'm not going to fall for that. We're staggered under the cases we're looking into now."

The House Committee on Science, Space and Technology, and its Subcommittee on Investigations and Oversight, chaired by Congressman Robert A. Roe has *not* been active in revealing wrongdoing in science. Yet those who have opposed Dingells' and Weiss's efforts are pleased with that committee. At a 1990 MIT conference on "Error, Fraud and Misconduct in Science," Dr. Mark Frankel, who heads the Office of Scientific Freedom and Responsibility (which keeps a file on science fraud) for the American Association for the Advancement of Science, noted that Roe's hearings into the general area of science fraud or misconduct had been "much more sympathetic to the scientific community" than were the hearings of either Dingell or Weiss. Roe's hearings, said Frankel, didn't "dwell on past instances of misconduct" but asked witnesses how scientists can improve matters. He pointed out that even the title of the Roe hearings, *Maintaining the Integrity of Scientific Research,* "takes a more positive approach" than the titles of the hearings of Dingell or Weiss. Their titles all contained the word "fraud," maintained Frankel. Frankel concluded that the differences between the hearings by Dingell and Weiss and those by Roe illustrate "some dispute" in Congress about "how much private versus public sector involvement" should exist in overseeing science.[19]

The Committee on Science, Space, and Technology has a long history of being liked by those it oversees, who themselves have been documented to be involved in dubious behavior. The previous chairman of this committee, when it was called simply the Science and Technology Committee was Congressman Don Fuqua. He went through one of the biggest and fastest revolving doors in American political history. On January 1, 1987, two days *before* leaving Congress (January 3, 1987) and the chairmanship of this committee, he took a job as president and general manager of the Aerospace Industries Association of America (AIA). Mr. Fuqua's official biography from the AIA describes him "as a leading spokesman for the U.S. aerospace industry," one of the major groups he had been allegedly overseeing. As his official biography states, "As Chairman of the Science and Technology Committee, Mr. Fuqua personally inspected and reviewed research projects . . . as far ranging as the development of competitive aircraft for the future . . . [and] the application of space technology. . . . The committee has direct authorization and oversight responsibility for most of the nation's government-funded civilian research, development, and demonstration programs."

WHY THE ALMOST INEVITABLE KANGAROO COURTS?

In every case examined in this book, the whistle-blower has suffered at least as badly and almost always far worse than the scientist who committed fraud.

It is a mistake to think that the administrative hearings faced by science whistle-blowers differ from the administrative hearings faced by whistle-blowers in other fields financed by federal money. Overwhelming documentation to prove the process follows the same pattern comes from the U.S. Civil Service System, some of whose members are scientists. The Civil Service Reform Act of 1978 set up a complicated system ostensibly to protect the rights of federal workers, but a 1978 Supreme Court decision, *Bush v. Lucas,* denied federal workers the right to sue, until they have exhausted all administrative procedures without success. Under this law, a civil servant who feels he or she has suffered retaliation for exposing wrongdoings can file a complaint with the Office of the Special Counsel. If the Special Counsel takes the case (which initially happened in only 1% of such filings), he will investigate the complaint before an administrative tribunal, the Merit System Protection Board.

A 1985 GAO survey discovered that although civil servants filed 11,000 complaints between 1979 and 1985, the Special Counsel failed to win a single one of these cases. He did negotiate 25 settlements. Of these, exactly 6 involved what the GAO euphemistically called "corrective action settlements."[20] The law has since been changed somewhat to protect whistle-blowers to a greater degree, but whether it will work in practice remains to be seen. Civil service whistle-blowers still do not have access to the federal courts.

Why do whistle-blowers in science and other fields come to grief? After analyzing a number of Pentagon cases, GAO investigator Franck C. Conahan explained: ". . . the reward system in place goes to those people who are successful in fielding weapons systems, and all support systems over there are in support of that sort of a philosophy.

"A person who gets this system fielded, or the group of people who get their system fielded, seem to enjoy a rather good series of rewards. You don't see that happening [to] people who get programs killed or stopped."[21]

Whistle-blowers inevitable have the effect of getting programs killed, stopped, slowed, or viewed with reduced expectations. During a telephone interview with me on June 21, 1991, Stanford University associate professor, science whistle-blower, and plaintiff's attorney Eugene Dong, MD, said, "The NIH is an agency that hands out money. That's what they do. They aren't too interested in seeing anything that diminishes that ability. The main problem I see currently is we have an agency/client relationship between the government funding agency and the universities. The more money the government agency gives out, the more power they get. That's how they get their power. Any actions which decrease what they hand out reduce the power of these officials."

Ernest Fitzgerald has thought long and hard about how money is handed out in Washington. His testimony on the subject is worth thinking about: "Washington, D.C. is essentially a one-industry town. Anyone who grew up, as I did, in a one-industry town or a mill town as they used to be called, can readily understand the seemingly complex goings-on in Washington if they give it a little thought.

"Our mill town's industry is not steel or textiles or some other useful product. Our industry is politics, and the lifeblood of politics is patronage. . . .

"Although they don't talk about it publicly, both the legislative and the executive branches are preoccupied with the distribution of patronage in all its forms. Because this distribution is so complex, both branches prize employees who are "responsive" in distributing the patronage as desired by the dominant party in the process.

"Because the patronage distribution process is sometimes messy, and not seen as a fit subject for full disclosure to the taxpayers and consumers, the legislative and executive branches have banded together to produce rules which govern and steer the behavior of a class of civilian employees called, I believe, 'inferior officers' in the Constitution, who do the day-to-day dirty work. These rules are the civil service laws and regulations."[22]

The National Science Foundation (NSF), the National Institutes of Health (NIH), the Department of Energy (DOE), and, the Department of Defense (DOD) are simply departments or agencies of the U.S. government. At the head of each one is a presidential appointee approved by the Senate. Beneath these presidential appointees are pyramids of "inferior officers"—civil servants. Each department or agency has a special mission: defending the country (DOD), researching and building nuclear weapons (DOE), advancing medical research (NIH), and advancing miscellaneous other areas of science (NSF). At their core, however, they all hand out money, much of it in the form of specific contracts, because although all of them make a few in-house products, they buy most of their products from outside suppliers. The Pentagon contractors receiving money refer to themselves as the "contractor community"; the scientists receiving money refer to themselves as the "scientific community." Often the only thing that members of these so-called communities have in common is that they get material sums of money from a limited number of federal agencies. The agencies, in turn, are supposed to rigorously oversee the scientific investigations although they rarely do so.

Erdem Cantekin, Margot O'Toole, Robert Sprague, Jon Kalb, and Roy Woodruff all pointed out irregularities that could have greatly embarrassed assorted officials at funding agencies and reduced their ability to give away money, which is the source of their power. Many other people in each field received funds from these officials or were colleagues of scientists who did. Few of the individuals involved have been eager to see any significant changes in how the money is handed out.

In short, science paid for by the federal government works exactly the same way as everything else paid for by the federal government—through the patronage system. Recognizing this simple fact makes all the sanctimony about peer review understandable; "peer review" is only a detail in running a patronage system. What is of primary importance is protecting or enlarging the system itself.

Congressman Dingell made an interesting observation in this regard: "The subcommittee has spent years investigating corruption in the defense industry and in other government agencies at all levels as well as industries which are subject to Federal regulation. . . . However, the scientific community, which has apparently been treated as a sacred cow for rather long, acts as if the government is making it a victim of some kind of persecution if it is questioned about accountability for Federal funds. . . . The subcommittee has reason to think that scientists are no different, no better, no worse than any other citizen. And that they can be expected to adhere to the same laws as all other American citizens. . . ."[23]

SOLUTIONS

The three solutions this book endorses for the problems illustrated about involve correcting the patronage system itself. With the explosion in scientific activity, the patronage system increasingly just is not up to the task. The greater the activity, as in Pentagon weapons development, super science projects, or the development of new drugs, the bigger the patronage system fiascos—the B1-B, the Laser Crosslink, the Space Shuttle, the space telescope, the X-Ray Laser of Star Wars, the Earthquake Engineering Research Center, and the amoxicillin tests. The Superconducting Supercollider is clearly part of the patronage system. Whether it will join the list of disasters remains to be seen, although many expect it will.

Separation of Funding and Control

Ernest Fitzgerald is the author of the first solution. He says that those who control the spending at funding organizations should never report in any way or be beholden in any way to those who get the money for the organization. This commonsensical separation of controls from spending has been muddied in the past few years in the Air Force and other Pentagon organizations. As Fitzgerald testified: "The primary function, as I have viewed it for many years, of a comptroller in the Department of Defense, and especially in the Air Force, has been to get money. He is the treasurer—it would be a better name for it. . . . I think I could say that of most of our military comptrollers. They do a good job of getting money. But controlling the money after it is gotten, and instituting management controls over the subordinate organizations is something that is hardly done at all."[24]

One immediate way to control wrongdoing is for Congress to repeal the case law created by *Bush v. Lucas,* in which the courts interpreted the Civil Service Reform Act to exclude civil servants' access to the federal courts. This would at least give civil service whistle-blowers the same legal rights as anyone who does not work for the federal government as a civil servant.

At the university level, university presidents, of course, help obtain money for the university. So when they exercise appointment power or review power over misconduct investigation panels that could result in the university getting less money, or giving back money to the federal government, they have a clear conflict. The same applies at lower levels to anyone on the inquiry or investigation panels who may also be part of a funding peer review panel. At the University of Pittsburgh, members of the English Department did a far better job of protecting Erdem Cantekin than did anyone from the Medical School.

The same separation between funding and control should apply to federal organizations such as the NIH or NSF. This became an issue in the Baltimore case, as Congressman Dingell explained at an August 1, 1991, hearing: "Dr. Healy [NIH Director] reportedly complained about the draft report of the Baltimore case, including words of praise for Dr. O'Toole but also refused to read [the] draft report, claiming that she could not bring herself to read the Baltimore report.

"Dr. Healy then ordered that Dr. Hadley [author of the Baltimore case report] should be reined in, and directed that Dr. Hadley should make no more decisions on the Gallo and Baltimore cases and further directed that all of Dr. Hadley's files be immediately removed from her office. Dr. Hadley, with her integrity impugned and her judgment in question by the head of the Agency and with an investigation against her being led by her boss at OSI [Office of Scientific Integrity], then resigned from further work on the cases."[25]

The need to separate control from funding means that the whole concept of using scientist "peers" to investigate allegations of misconduct is inherently a bad idea. They remain peers only of the accused, who at the time of the so-called investigations (and usually even afterward) stay in the peer-reviewed patronage system. As case after case has shown, whistle-blowers always or nearly always become excommunicants.

Conflicts of Interest in Medical Research

Congressman Ted Weiss has offered a second solution, limited primarily to medical research conducted at universities. Weiss introduced a bill, H.R. 1819,[26] to the Committee on Energy and Commerce, chaired by John Dingell. This legislation would encourage universities receiving federal research money to prevent or at least publicize conflicts of interest. The proposed law would allow the Secretary of Health and Human Services, who ultimately is in charge of the NIH, to take money back from universities if the NIH disapproves of perceived conflicts of interest. This could be an effective sanction. Right now, there are no sanctions against conflict of interest in federally funded university research.

The Secretary would have the option to (1) require the university to disclose the conflict of interest "in each public presentation of the results of the research," (2) do anything else the Secretary "considers to be appropriate" to protect the public interest, (3) cut off the money for the project, or (4) take all or part of the money back from the university for the period "in which the conflict exists."

Furthermore, the law specifies that unless the university certifies annually that it is obeying the regulations, the Secretary must lock them out of the money. Conflict of interest does not have to be linked to improper behavior; in the proposed legislation the conflict of interest *is* improper behavior.

Another bill, H.R. 2507, is intended to provide protection to whistle-blowers against retaliation from their universities. The bill requires institutions applying for Public Health Service funds to set up procedures and policies to safeguard whistle-blowers or anyone else who helps in an investigation. The bill also requires the HHS Secretary to oversee the institutions to make sure they live up to their procedures. The universities would lose money if their investigations into allegations of misconduct were not accurate, thorough, or fair.

The proposed legislation would also guarantee access to other researchers' data at a reasonable cost, after the initial publication of the results in a scientific journal or three years after the federal money for the project has ended. It would also require the Secretary to "establish recommendations regarding timely dissemination of any . . . results that have any clinical application to a disease or disorder that poses a significant threat to the public health."

The legislation has certain gaps. First, it excludes all areas of science except applied medical research. Second, although the university may feel the pain of a sanction, the whistle-blower may never enjoy the pleasure of a remedy. Under this or probably any other conceivable reform, the whistle-blower will still likely become an excommunicant.

The Federal False Claims Act

A third solution applies to all areas of federally supported science, provides sanctions against the offenders, remedies to the whistle-blowers, and due process to everyone. The legislation already exits, but it needs further legislative refinement. In this scenario, the whistle-blower simply sues under the *qui tam* provisions of the Federal False Claims Act. The possible targets of the suits are (1) the offending university, (2) the person who the whistle-blower thinks has defrauded the government by compromising the science, and possibly (3) the appropriate federal officials who may have helped to cover up improper behavior.

Congress passed the Federal False Claims Act (sometimes referred to as the "Abraham Lincoln Law") during the Civil War to prosecute war profiteers. Congress updated the law in 1986 with the idea of cracking down on Pentagon contractors.

A whistle-blower who files suit is able to receive due process in the federal courts without going through university procedures. "The False Claims Act specifically doesn't provide for or require any administrative proceedings. You can go *directly* to the Federal courts," says Eugene Dong, MD, JD, an associate professor in Stanford's Department of Cardiovascular Surgery and the author or coauthor of more than 100 articles dealing with heart research.

The case has to involve fraud against the U.S. government, but as Dong notes: "Ninety-five percent of scientific activity in the U.S. is funded by the U.S. government. Therefore, a scientific fraud is actually a fraud against the U.S. government and directly against the people of the U.S. It's no longer then a mere unethical activity which will be settled by gentlemanly discourse amongst scientists."

Dong, acting in his capacity as attorney, is now representing the first whistle-blower to sue under this law. "The Act takes the case out of the hands of the OSI, out of the hands of the universities and puts it into the courts," said Dong.

There is one catch; for the first three months after a complaint is filed, it is kept under seal. During this time, the Justice Department has an opportunity to decide if it wants to join in the suit. Naturally, the Department first goes to the sister federal agency that originally funded the grant. In the case of Dong's suit, this is the NIH, which then consults its own OSI, for guidance. According to Suzanne Hadley, who wrote the OSI draft reports on the Bluestone and Baltimore cases, "They [the Justice Department] would say, 'You examine the scientific issues here, and you tell us if the government should intervene in this case.' And, when this has happened, we convene a body of scientists." So the case ends up in front of the patronage "peers" anyway.

Dong explained how he handled this built-in sand trap. "You have to go and bash their experts and make sure they don't have a conflict of interest. I went personally and I insisted on going with my client. OSI said they wanted to interview my client. I said *not* without Justice Department lawyers present."

According to Dong, the OSI meetings included not just the OSI panelists and staff but also the plaintiff, the Justice Department lawyers, and Dong himself. "We were arguing our case and teaching the Justice Department at the same time," said Dong, adding that the plaintiff "had evidentiary material, xeroxed copies of notebooks, which we went through carefully. It took eight months." By taking this approach and forcing some semblance of due process into the OSI procedure, Dong's was successful. The Justice Department joined his suit.

Is it easier for a whistle-blower to win by this approach than the administrative one? "All you have to show is that the defendant knew the consequences even if he didn't intend them. Furthermore, the burden of proof is 'the preponderance of the evidence,' which if you were to put a number on it means roughly 50.1% of the evidence," said scientist/lawyer Dong.

The Abraham Lincoln Law/Federal False Claims Act does provide for a remedy if the whistle-blower wins. "The current law is a nice remedy," said Dong. The defendant is fined with *treble* damages, and the private plaintiff, or "relater" in the language of the statute, receives a minimum of 10% and a maximum of 25% of the treble damages plus $5,000 to $10,000 for each of the defendant's offenses. Those are the rewards if the Justice Department chooses to join the suit. If the Justice Department does not join the suit, the plaintiff gets 30% of the treble damages, plus attorney fees from the defendants.

"In a science fraud case, the whole grant amount should be considered the damage. Had the granting agency been aware of any fraud, they wouldn't have issued the grant," says Dong. In the case he is prosecuting, Dong estimates the damages to the government at about $1.2 million. Treble damages could bring that to $3.6 million plus whatever the dollar amount is for the specific offenses.

The *qui tam* provisions of the Federal False Claims Act also provide protection against retaliation to the whistle-blower, but this remedy is highly imperfect. Explained Dong, "Any person in the employment of the defendant who has had some sort of job action taken against him *for the reason* of the knowledge which he is relating can file a lawsuit under the Act, and can get personal damages, but cannot get triple damages on the personal claim. But he can still get 25% of the relater claim itself." The complicating factor is that it is difficult to show why someone was fired, demoted, or passed over. It requires proving a motive or state of mind, which is almost impossible to do—unless the retaliator confesses.

Congress should adopt the most recent change in the Civil Service Reform Act for *qui tam* suits as well. Under the current version of the Civil Service law, the whistle-blower doesn't have to establish the retaliator's motive. The whistle-blower merely has to show that the truth telling was a "contributing factor" in the job action. This has been interpreted to mean, "any factor which alone or in connection with other factors, tends to affect in any way the outcome of the decision."[27]

An argument sometimes made against *qui tam* suits is that they provide a monetary incentive to become a whistle-blower and that this may in some way reduce their credibility in the minds of the public. However, the credibility of the whistle-blower in the public mind does not change the improper conduct of the institutions. Also, rewarding whistle-blowers and their lawyers may be effective in creating a self-regulating, systemic mechanism for dealing with fraud cases. Dong believes there are plenty of lawyers right now who could handle the cases. "For example," he says, "patent lawyers are scientifically trained."

Dong says of his test case, which as of this writing has not yet gone to trial, "Somebody had to break the ice, and we did it."

This approach is significant for what it does not do. It does not create another federal bureaucracy, a "science police." It does not require "data audits." Nevertheless, it *may* force scientists, on their own, or with the encouragement of their universities, to live up to the strictures of the scientific method.

Notes

Introduction

1. Daniel E. Koshland, "Fraud in Science," *Science* 235 (9 January 1987): 141.

2. William Broad and Nicholas Wade, *Betrayers of the Truth: Fraud and Deceit in the Halls of Science* (New York: Touchstone, 1983), 87.

3. *Scientific Fraud and Misconduct and the Federal Response: Hearing before the Human Resources and Intergovernmental Relations Subcommittee of the Committee on Governmental Operations,* 11 April 1988, 108.

4. Ibid., testimony of June Price Tangney, 98–102.

5. *Washington Post,* 13 July 1989, p. A16.

6. U.S. Congress, Office of Technology Assessment, *Federally Funded Research: Decisions for a Decade,* OTA-SET-490 (Washington, DC: U.S. Government Printing Office, May 1991), 6–7.

7. Ibid., 249.

8. Frank Solomon, "Error, Fraud and Misconduct in Science," Conference, Massachusetts Institute of Technology, 7 April 1990.

9. U.S. Congress, *Federally Funded Research,* 197.

10. *Apparent Financial Wrongdoing by an Official in the Laboratory of Tumor Cell Biology at the National Institutes of Health, Hearing before the Subcommittee on Oversight and Investigations of the House Committee on Energy and Commerce,* 6 March 1991, 2–9.

Chapter 1

1. Charles H. Herz, General Counsel, NSF, letter to Eric R. Glitzenstein, Public Citizen, 12 March 1990, 1.

2. Jon Kalb, *A "Covert" Process or Due Process?, The Peer Review Process of the U.S. National Science Foundation,* additional material submitted for the record; prepared for the Task Force on Science Policy of the House Committee on Science and Technology, 99th Cong., 2nd sess., 8–10 April 1986.

3. John Conlan, quoted in *University Funding Information on the Role of Peer Review at the NSF and NIH,* General Accounting Office, GAO/RCED-87-87FS, March 1987, 7.

4. "NSF Director's Statement on the Report of the Advisory Committee on Merit Review," *Final Report on Merit Review,* NSF Advisory Committee on Merit Review, (Washington, DC: National Science Foundation, 1986), NSF 86-93.

5. "Proposal Review at NSF: Perceptions of Principal Investigators," *Report of a Survey by NSF's Program Evaluation Staff,* Report 88-4 (Rev. 4-90) (Washington, DC: National Science Foundation, February 1988), Appendix.

6. Kalb, *"Covert" Process or Due Process?*

7. Eliot Marshall, "Gossip and Peer Review at NSF," *Science* 238 (11 December 1987) 1502.

8. Jon Kalb, *A History of U.S. National Science Foundation Sponsored Prehistory Research in Ethiopia, 1973–1983—Questions and Issues,* unpublished manuscript (1986), endnote 60.

9. Iwao Ishino, Program Director for Anthropology to General Counsel, NSF, Memorandum, 19 September 1974.

10. Kalb, *A History,* endnote 60.

11. Donald Johanson, letter to Dr. Iwao Ishino, NSF, undated (approximately dated by Kalb, who obtained it under a FOIA request, as September 13, 1974).

12. John E. Yellen, *Jon Kalb Reconsideration,* Memorandum (hand-dated "27 XII '77"), 4.

13. Kalb, *A History,* 19.

14. *Science* 219 (14 Jan. 1983) 149.

15. Marina and David Ottaway, *Ethiopia—Empire in Revolution,* Africana Publishing Company, N.Y., 1978, p. 11.

16. Yellen, *Jon Kalb,* 3.

17. Complaint, Kalb v. National Science Foundation, ¶ 10: 4.

18. "Conflict-of Interest Statement for NSF Advisory Committee/Review Panel Members," NSF Form 1230 (10-88).

19. Complaint, Kalb v. National Science Foundation, ¶ 11: 4.

20. Memorandum of Points and Authorities in Support of Motion to Dismiss Or, In the Alternative, For Summary Judgment, Kalb v. National Science Foundation, D.D.C., Civ. No. 86-3557.

21. Memorandum of Points and Authorities.

22. Stipulation of Settlement, Kalb v. National Science Foundation, U.S., D.D.C. Civ. No. 86-3557, 8 December 1987, ¶ 2c.

23. James L. Bostick, NSF Grants Officer, letter to Chancellor, University of California, Berkeley, re. Grant No. BNS76-14333 (7 May 1976).

24. Stipulation of Settlement, ¶ 2b.

25. Ibid., ¶ 2c.

26. Ibid., ¶ 2b.

27. Referenced in letter to Kalb of 6 December 1983, from the NSF Director of the Office of Legislative and Public Affairs. Item 7 in this letter reads as follows: "The notes made by the General Counsel's Office pertaining to telephone conversations with reviewers are enclosed." The enclosure is a photocopy of the handwritten notes on ruled legal paper, stamped with the FOIA document number.

28. Charles Redman, *Report on Jon Kalb's Case,* 29 October 1982.

29. Nancie Gonzalez declined to be interviewed for this book.

30. Complaint, Kalb v. National Science Foundation, ¶ 13.

31. Ibid., ¶ 18, 6.

32. Proposal Rating Sheet for Proposal DEP-7717065, Title of Proposal: *Paleoecological investigations in the Awash Valley, Afar Depression, Ethiopia,* 2.

33. Ibid., 1.

34. Redman, *Report,* item 6.

35. John Yellen, letter to Desmond Clark, December 7, 1977 (FOIA document number 00423).

36. *Hearings before the Task Force on Science Policy of the House Committee on Science and Technology,* 99th Cong., 2nd sess., 8–10 April 1986, No. 134, 634.

37. Stipulation of Settlement, ¶ 2d.

38. Ibid., ¶ 2f.

39. Memorandum of Points and Authorities, 3.

40. Ibid., 2–3.

41. Ibid.

42. Redman, *Report,* items 3 and 5.

43. Ibid., item 5.

44. Complaint, Kalb v. National Science Foundation, ¶ 8.

45. *Science* 238 (11 December 1987) 1502.

46. *Science Policy Study—Hearing Volume 17, Research Project Selection: Hearings before the Task Force on Science Policy of the House Committee on Science and Technology,* 99th Cong., 2nd sess., 8–10 April 1986, No. 134, 636.

47. John Yellen, *Report on Paris and Addis Ababa Conferences Concerning Future Research in the Ethiopian Rift Valley,* undated, 6.

48. "All Things Considered," National Public Radio, CIA Ethiopia, No. 821208-9, Tape.

49. Kalb, *A History,* 12.

50. Redman, *Report.*

51. Ibid., item 6.

52. Jerome Fregeau, *In the Matter of Mr. Jon E. Kalb,* Memorandum to File from NSF Director of Audit and Oversight, 19 November 1982.

53. Complaint, Kalb v. National Science Foundation, ¶ 26: 9.

54. Memorandum of Points and Authorities, 18.

55. *Electronic Records: Legal and Policy Considerations,* Report of the Electronic Records Committee, National Science Foundation, Washington, DC, April 1987, 7–8.

56. "NSF agrees to make peer review process for research grants open and accountable," press release from Public Citizen, 14 March 1990.

57. Daryl Chubin, "Proposal Pressure in the 1980s: An Indicator of Stress on the Federal Research System," U.S. Congress, Office of Technology Assessment, Washington, DC. The figures are based on Table 2.

58. Office of Inspector General, *Semiannual Report to the Congress,* Number 2, 31 March 1990, 15.

59. Ibid., 15.

Chapter 2

1. *NSF Authorizations, Hearings before the Senate Committee on Commerce, Science and Transportation,* 100th Cong., 2nd Sess., 14 April 1988, 18.

2. *Science* 250: 620.

3. For *peer review,* the NSF claims that "the judgment of scientific excellence lies at the heart of the funding decision." Four other criteria enter into peer review, listed here verbatim from an NSF report on merit review: (1) research performance competence, (2) intrinsic merit of the research, (3) utility of relevance of the research, (4) effects of the research on the infrastructure of science and engineering. *Merit review* includes a few other considerations as described by the NSF's *Final Report on Merit Review:* "In the engineering Research Centers Program . . . first stage peer panels judge the technical excellence of a proposed center and the impact of its educational plan and its plan to involve industry. Then a more broadly based panel is convened to review the best of the proposals for technical and managerial excellence and the importance of their topics to national productivity. Technical excellence and the importance of the topic for technological competitiveness are paramount in the deliberations; geographic distribution and the impact on engineering infrastructure are additional criteria." *Final Report on Merit Review,* NSF Advisory Committee on Merit Review, (Washington, DC: National Science Foundation, 1986) NSF-86-93, 14–16.

4. Committee on Earthquake Engineering, National Research Council, *Review of Earthquake Engineering Research in the National Earthquake Hazards Reduction Program,* (Washington, DC, 11 April 1990), V.

5. U.S. General Accounting Office, "NSF Problems Found in Decision Process for Awarding Earthquake Center," GAO/RCED-87-146, 10 [hereinafter GAO Report].

6. Ibid., 27.

7. Ibid., 26.

8. Ibid., 27.

9. *NSF Authorizations,* 21.

10. Ibid., 21.

11. David W. Cheney, *The National Earthquake Hazards Reduction Program,* CRS Report for Congress (9 August 1989), 89-473 SPR, 122.

12. *Congressional Record* (October 6, 1986), S 15200-15201.

13. Cheney, *National Earthquake Hazards,* 8.

14. National Science Board, closed door meeting, 27 (verbatim transcript).

15. Interview with California proposal principal, who requested anonymity, in his office, 17 November 1990.

16. *GAO Report,* 18–20.

17. *GAO Report,* 42.

18. Memorandum from Jerome H. Fregeau, Director, Division of Audit and Oversight, via Controller to Director, 10 November 1986, 3.

19. Earthquake Engineering Research Center, *Site Visit to the University at Buffalo, State University of New York, July 8–9, 1986,* 5.

20. *Reasons for Questioning the Propriety of the NSF Decision to Award the Earthquake Engineering Research Center to SUNY–Buffalo,* item 7: 4. The NSF's "Memorandum to Members of the National Science Board" (August 13, 1986) stated: "The review conducted for the Earthquake Engineering Research Center was carried out by a Blue Ribbon Panel. . . . The site visit team consisted of the seven Blue Ribbon Panel members and six NSF staff members. . . . The site visit panel judged the Center as highly meritorious and the Center was recommended for funding."

21. National Science Board, closed door meeting, 28 (verbatim transcript).

22. *Reasons for Questioning Propriety,* item 2: 1.

23. Ibid.

24. Robert Olson et al., *To Save Lives and Protect Property—An Interpretive Review of Federal Earthquake Activities, 1964–1987,* prepared for the Federal Emergency Management Agency (FEMA) by VSP Associates, Inc. (22 December 1987), 120.

25. "Aftershock of an Earthquake Study Grant," *Washington Post,* 2 October 1986, p. A7.

26. GAO Report, 61.

27. *NSF Authorizations,* 21.

28. GAO Report, 25.

29. Earthquake Engineering Research Center, *Site Visit to the University of California at Berkeley, August 9, 1986,* 2.

30. National Science Board, closed door meeting, 29 (verbatim transcript).

31. *Reasons for Questioning Propriety,* 5.

32. National Science Board, closed door meeting, 23 (verbatim transcript). Also quoted, with correction, in David Willman, "Wilson Alleges Age Bias Denied Quake Grant to California," *San Jose Mercury News,* 31 October 1986, 5A.

33. GAO Report, 51.

34. Ibid., 50.

35. Earthquake Engineering Research Center, *Site Visit to the University of Buffalo, State University of New York, July 8–9, 1986,* 6.

36. GAO Report, 49.

37. Ibid., 54.

38. Ibid., 55.

39. Ibid., 48.

40. Ibid., 28.

41. Ibid.

42. Ibid., 50.

43. "Loss of Earthquake Study Center Shakes Up California Scientists," *Los Angeles Times,* 23 October 1986, Part I: 3.

44. Olson, et al., *To Save Lives and Protect Property,* 124.

45. GAO Report, 41.

46. National Science Board, closed door meeting, 24 (verbatim in transcript).

47. Thomas S. Bateman and Carl P. Zeithaml, *Management* (Homewood, IL: Irwin, 1990), 490.

48. Arthur G. Bedeian, *Management,* 2nd ed. (Chicago: Dryden Press, 1989), 436.

49. The passage also included the highly debatable point, "The Center concept, conversely, required 'leadership' or 'management' of a type unfamiliar to the Californians." Olson, et al., *To Save Lives and Protect Property,* 122.

50. GAO Report, 41.

51. R. Wayne Mondy and Robert M. Noe, III, *Personnel,* 3rd ed., (Boston: Allyn and Bacon, 1987), 220.

52. GAO Report, 46.

53. Ibid.

54. National Science Board, closed door meeting, 32 (verbatim transcript).

55. *Reasons for Questioning Propriety,* item 11: 5–6.

56. *A Report to the National Science Foundation on the Third Year Site Evaluation of the National Center for Earthquake Engineering Research,* (7 August 1989), 10.

57. Olson et al., *To Save Lives and Protect Property,* 126.

58. GAO Report, 31.

59. Reasons for Questioning Propriety, item 5: 3.

60. Earthquake Engineering Research Center, *Site Visit to the University of California at Berkeley, August 9, 1986,* 4–5.

61. GAO Report, 48.

62. Ibid.

63. *The Buffalo News,* 6 April 1988, A-12.

64. GAO Report, 47.

65. Ibid., 33.

66. Ibid., 47.

67. Ibid., 11.

68. Ibid., 28.

69. Ibid., 35.

70. Ibid., 22.

71. Ibid.

72. *NSF Authorizations,* 4.

73. Wilfred Iwan, quoted in "Loss of Earthquake Study Center Shakes Up California Scientists," *Los Angeles Times,* 23 October 1986, Part I: 3.

74. National Science Board, *Science & Engineering Indicators—1989,* NSB-1 (Washington, DC: U.S. Government Printing Office, 1989), 272.

75. *NSF Authorizations,* 43.

76. National Science Board closed door meeting, 31.

77. *Statement before the Senate Committee on Commerce, Science and Transportation,* Thursday, 14 April 1988, 7.

78. Committee on Earthquake Engineering, National Research Council, *Review,* 37.

79. Ibid., 37–38.

80. "Loss of Earthquake Study" Part I: 3.

81. Cheney, *National Earthquake Hazards,* 130.

82. Committee on Earthquake Engineering, National Research Council, *Review,* 38.

83. Ibid., 31.

84. Ibid., 35.

85. *Report on Third Year Site Evaluation,* 2.

86. Committee on Earthquake Engineering, National Research Council, *Review,* 33.

87. "Loss of Earthquake Study Center," 3.

88. *Report on Third Year Site Evaluation,* 14.

Chapter 3

1. Christopher T. Hill, "Considerations in Funding Large-Scale Science," *CRS Review* (February 1988), 3.

2. William C. Boesman, "Superconducting Super Collider," *CRS Review* (February 1988), 18.

3. Claudine Schneider, "The Superconducting Supercollider: A Collision Course," *The Christian Science Monitor,* 23 November 1987.

4. Michael Barone and Grant Ujifusa, *The Almanac of American Politics* (1990) 1172.

5. U.S. General Accounting Office, "Site Selection Process for the Department of Energy's Super Collider," GAO/T-RCED-89-22, 1 (testimony of Flora H. Milans).

6. The entire story is reported by Phil Kuntz in "Pie in the Sky: Big Science Is Ready for Blastoff," *Congressional Quarterly Weekly Review,* 28 April 1990.

7. Irene Stith-Coleman, "Proposal to Map and Sequence the Human Genome," *CRS Review* (21 April 1989), 10.

8. Robert K. Moyzis, Director, Center for Human Genome Studies, Los Alamos National Laboratory, Los Alamos, New Mexico: Statement before the Subcommittee on Energy Research and Development of the Senate Committee on Energy and Natural Resources, 11 July 1990, 7–8.

9. "SDI Contracts Data Base Information, Contract Dollars By State," SDIO, 29 January 1990.

10. Memorandum for General Watts, SAF/AC, 16 May 1988, 3.

11. *Hearings before the Subcommittee on Science, Technology and Space, Senate Committee on Commerce, Science and Transportation, Oversight Hearing on the Status of the Hubble Space Telescope* (stenographic transcript), 29 June 1990, 20–21 (quoted by Senator Gore).

12. Philip M. Boffey, *The New York Times,* 22 February 1986.

13. *Washington Post,* 23 June 1987.

14. NASA, Office of Inspector General, *Semiannual Report,* (1 April 1989–30 September 1989).

15. *Hearings before the Subcommittee on Science, Technology and Space,* 89.

16. Ibid., 120.

17. William J. Broad, "Hubble Has Backup Mirror, Unused," *The New York Times,* 18 July 1990.

18. *Hearings before the Subcommittee on Science, Technology and Space,* 46.

19. Ibid., 37.

20. U.S. General Accounting Office, "Federal Research—Super Collider Estimates and Germany's Industrially Produced Magnets," GAO/RCED-91-94FS, 11–12.

21. Alex Roland, "The Shuttle—Triumph or turkey?" *Discover* (November 1985), 48.

22. *Hearings before a Subcommittee of the House Committee on Appropriations,* 101st Cong., 2nd Sess., 20 March 1990, 5.

23. Report to the Congress by the Comptroller General of the United States, *NASA Should Provide the Congress Complete Cost Information on the Space Telescope Program,* 3 January 1980, 6.

24. Ibid., 9.

25. U.S. General Accounting Office, "Space Science, Status of the Hubble Space Telescope Program," GAO/NSIAD-88-118BR, May 1988, 17.

26. *Hearings before the Subcommittee on Science, Technology and Space,* 36.

27. Senate Armed Services Committee, *The President's Strategic Defense Initiative,* Testimony submitted for the record for the Senate Armed Services Committee, 24 April 1984 (quoted by Richard L. Garwin).

28. Secretary of Energy John Herrington's press conference, 22 July 1988 (verbatim transcript).

29. Clarence A. Robinson, Jr., "Advance Made on High-Energy Laser," *Aviation Week & Space Technology* (23 February 1981).

30. Deborah Blum, "Father of the H-Bomb: X-Ray Laser Unproven," *Sacramento Bee,* 10 December 1987, A-1.

31. Quoted in D. M. Ritson, "A Weapon for the Twenty-First Century," *Nature* 328 (6 August 1987): 487.

32. Edward Teller, "SDI: The Last, Best Hope," *Strategic Defense Initiative,* 28 October 1985, 75.

33. Joseph Romm (with advice from Joseph Cirincione), "Pseudo-Science and SDI," *Arms Control Today,* October 1989, 17.

34. Ibid., 17.

35. U.S. General Accounting Office, "Strategic Defense Initiative Program—Accuracy of Statements Concerning DOE's X-Ray Laser Research Program," GAO/NSIAD-88-181BR, 4.

36. Ibid., 5.

37. Harold Brown, "Is SDI Technically Feasible?" *Foreign Affairs,* TK, 1985, 442.

38. Judy England-Joseph, Associate Director, GAO, Energy Issues, *Concerns about Developing and Producing Magnets for the Superconducting Super Collider, Testimony before the Subcommittee on Investigations and Oversight, House Committee on Science, Space and Technology,* May 9, 1991, GAO/T-RCED-91-51, 9.

39. Quoted in William J. Broad, "Crown Jewel of 'Star Wars' Has Lost Its Luster," *The New York Times,* 13 February 1990.

40. Ibid.

41. GAO, *Strategic Defense,* 5.

42. *Weekly Bulletin,* Wednesday, 18 December 1985, 3.

43. U.S. General Accounting Office, "SDI Program—Evaluation of DOE's Answers to Questions on X-Ray Laser Experiments," GAO/NSIAD-86-140BR.

44. Liz Mullen, "Physicist Teller Speaks at UCI on Space Arms Race," *Los Angeles Times,* Orange County ed., 4 April 1985.

45. *Classification Bulletin,* signed by Herman H. Teifeld, Classification Officer, Lawrence Livermore National Laboratory, 10 May 1985.

46. Robert Scheer, "The Man Who Blew the Whistle on 'Star Wars,'" *Los Angeles Times Magazine,* 17 July 1988.

47. Secretary of Energy John Harrington's press conference, 22 July 1988 (verbatim transcript).

48. Roy Woodruff, Grievance letter to UC President Pierpont Gardner, 3 April 1987, 1–2.

49. GAO, *Strategic Defense,* 3.

50. Keith Rogers, "Lab Conflict Is Alleged in Hiring Prober in Laser Study," *Valley Times,* 19 April 1989, 1.

51. U.S. General Accounting Office, "DOD Revolving Door—Post DOD Employment May Raise Concerns," GAO/NSIAD-87-116, p. 1.

52. Vincent Kiernan, "Budget Cuts Stymie X-Ray Laser Tests," *Space News,* 13 November 1989.

53. Vincent Kiernan, "Excess Hype of X-Ray Laser Causes Funding GAP for Weapons Research," *Space News,* 9 October 1990.

54. Keith Rogers, "Livermore Lab Director Disputes Teller Allegations," *Valley Times,* 24 October 1987.

55. Discussed in a speech to the U.S. House of Representatives by Congressman George E. Brown, Jr., 14 July 1988. The actual California legislative language is in Item 6440-001-001, *Joint Legislative Conference Committee on the Budget,* 27 June 1988, kindly supplied to the author by the University of California.

Chapter 4

1. House Committee on Government Operations, *Are Scientific Misconduct and Conflicts of Interest Hazardous to Our Health? 19th Report together with dissenting and additional views, September 10, 1990,* (Washington, DC: U.S. Government Printing Office, 1990) [hereinafter *Are Scientific Conduct and Conflicts of Interest Hazardous?*] 12.

2. *Fraud in NIH Grant Programs: Hearing before the Subcommittee on Oversight and Investigations of the House Committee on Energy and Commerce,* 100th Cong., 2nd Sess., 12 April 1988, 159.

3. Ibid., 152.

4. *Scientific Fraud and Misconduct and the Federal Response:* Hearing before the Subcommittee of the House Committee on Government Operations, 100th Cong., 2nd Sess., 11 April 1988, 9.

5. Ibid.

6. Ibid., 23.

7. Ibid., 19–24 (reprint of letter to Natalie Reatig, 20 December 1983).

8. Ibid., 25–27 (reprint of letter to Dr. Thomas Detre, 17 January 1984).

9. Ibid., 32 (reprint of letter to Dean Donald Leon, MD, 17 February 1984).

10. Ibid., 36 (reprint of letter to Dr. Donald Leon, 8 May 1984).

11. Ibid., 38–39 (reprint of letter to Lorraine B. Torres, 6 July 1984).

12. Ibid., 12.

13. Ibid., 12–13.

14. Telephone interview with author, 23 June 1991. All quotes not otherwise cited are from this interview.

15. *Are Scientific Misconduct and Conflicts of Interest Hazardous?* 12.

16. Ibid., 63.

17. *Maintaining the Integrity of Scientific Research: Hearing before the Subcommittee on Investigations and Oversight of the House Committee on Science, Space, and Technology,* 101st Cong., 1st Sess., 28 June 1989, 200.

18. *Fraud in NIH Grant Programs,* 163.

19. *Maintaining the Integrity of Scientific Research,* 202 (reprint of letter to Robert L. Sprague, 28 July 1989).

20. Washington, DC interview with author, June 20, 1991.

21. *Fraud in NIH Grant Programs,* 159.

22. Ibid., 161.

23. *Are Scientific Misconduct and Conflicts of Interest Hazardous?* 14.

24. *Fraud in NIH Grant Programs,* 161.

25. Discussion after B. R. Kopp (Moderator), "Research Fraud from the Inside Out" (From a symposium at the Society of Research Administrators, Boston, 11 October 1988) (verbatim transcript).

26. "Error, Fraud and Misconduct in Science," The Arthur Miller Lecture on Science and Ethics, Day long conference, MIT, 7 April 1990.

27. William Broad and Nicholas Wade, *Betrayers of the Truth,* (New York: Simon & Schuster, 1983) 10.

28. *Are Scientific Misconduct and Conflicts of Interest Hazardous?* 11.

29. Weaver, D., Reis, M. H., Albanese, C., Constantini, F., Baltimore, D., and Imanishi-Kari, T. (1986). "Altered Repertoire of Endogenous Immunoglobulin Gene Expression in Transgenic Mice Containing a Rearranged mu Heavy Chain Gene." 45 *Cell:* 247–259.

30. *The New York Times,* 3 December 1991, 1.

31. *Scientific Fraud: Hearing before the Subcommittee on Oversight and Investigations of the House Committee on Energy and Commerce,* Stenographic Minutes, 9 May 1989, 59.

32. *Scientific Fraud,* 61.

33. *Scientific Fraud,* 204.

34. *Scientific Fraud,* 227–228.

35. Interview with Walter Stewart, May 5, 1989, in his "office" with Dingell's Oversight and Investigations Subcommittee. Stewart's office was a tiny alcove near the photocopier. All quotes from Stewart, unless otherwise indicated, are from this day-long interview.

36. David Baltimore, "Dear Colleague" letter, 19 May 1988, 7.

37. *Scientific Fraud,* 4 May 1989, 4–5.

38. Margot O'Toole, Statement, May 3, 1991.

39. Ibid.

40. Quoted in *Fraud in NIH Grant Programs,* 67.

41. *Fraud in NIH Grant Programs,* 95.

42. *Scientific Fraud,* Written testimony of Bridgette Huber, PhD, 9 May 1989, 3.

43. Tufts Inquiry Panel, Minutes of Ad-Hoc Committee Meeting, Brigitte Huber, PhD, Robert Woodland, PhD, Henry Wortis, MD, Chair, 4 June 1986.

44. *Scientific Fraud,* 9 May 1989, 90.

45. Ibid., Testimony of Dr. Henry Wortis submitted for the record, 9 May 1989, 6.

46. *Scientific Fraud,* Testimony of Dr. Brigitte Huber submitted for the record, 9 May 1989.

47. *Scientific Fraud,* Testimony of Dr. David Baltimore submitted for the record, 4 May 1989, 6.

48. Baltimore, "Dear Colleague" letter, 4–5.

49. *Fraud in NIH Grant Programs,* 98–99.

50. Ibid., 106–107.

51. David Baltimore, Confidential Memorandum to Herman Eisen, 9 September 1986.

52. *Scientific Fraud,* 9 May 1989, 186.

53. Walter W. Stewart and Ned Feder, Draft manuscript, 30 September 1987.

54. *Fraud in NIH Grant Programs,* 66–70.

55. Benjamin Lewin, Letter to Walter Stewart, 19 October 1987.

56. Benjamin Lewin, Letter to Walter Stewart, 15 December 1987.

57. Patricia A. Morgan, Letter to Walter Stewart, 16 December 1987.

58. *Scientific Fraud,* 4 May 1989, 203.

59. "Credibility," Editor's Page, *Chemical & Engineering News,* 22 May 1989.

60. Scientific Fraud, 4 May 1989, 220.

61. Office Of Scientific Integrity, Ref. 072, "Draft Investigative Report, Investigation Report Concerning the Weaver et al. 1986 *Cell* Paper," 14 March 1991, 7–8.

62. *Scientific Fraud,* 4 May 1989, 202.

63. "Baltimore Wins PR Battle, but Key Issues Remain," *Science & Government Report,* 15 May 1987, 7.

64. Philip Sharp, "Sample Letter to Congressmen, Editors," 1.

65. Stephen Jay Gould, "Judging the Perils of Official Hostility to Scientific Error," *The New York Times,* 30 July 1989, sec. 4: 6.

66. "Latest Chapter in the Fine Science of the Smear," *The Wall Street Journal,* 5 May 1989.

67. "M.I.T. to Repay $731,000 Including Lobbyist Fees," *The New York Times,* 24 April 1991.

68. *Scientific Fraud,* 9 May 1989, 30.

69. Ibid., 4 May 1989, 91–92.

70. Ibid., 9 May 1989, 30.

71. Ibid., 4 May 1989, 79.

72. Office Of Scientific Integrity, "Draft Investigative Report," 114–115.

73. "'I Am Innocent,' Embattled Biologist Says," *The New York Times,* 4 June 1991, C1.

74. Office Of Scientific Integrity, Draft Investigative Report, 119–120.

75. *Scientific Fraud,* 9 May 1989, 198.

76. *The New York Times,* 21 March 1991, A1.

77. David Baltimore, Comments to OSI, "Draft Investigative Report," 3 May 1991, 1.

78. "Margot O'Toole's Record of Events," 351 *Nature* (16 May 1991): 183.

Chapter 5

1. House Committee on Government Operations, *FDA's Deficient Regulation of the New Drug Versed: 71st Report together with Additional Views,* 100th Cong., 2nd Sess., House Report 100-1086, 13 October 1988, 40.

2. Devra Lee Davis, PhD, MPH., Statement before the Subcommittee on Courts and Administrative Practice, Senate Committee on the Judiciary, 17 May 1990, 3.

3. *FDA's Regulation of Zomax: Hearings before a Subcommittee of the House Committee on Government Operations,* 98th Cong., 1st Sess., 26–27 April 1983, 15–16.

4. Stephen Alexander, MD, Testimony, *FDA's Regulation of Zomax: Hearings,* 1983, 10.

5. *FDA's Regulation of Zomax: Hearings,* 30.

6. Ibid., 10.

7. House Committee on Government Operations, *FDA's Regulation of Zomax: 31st Report together with Additional and Dissenting Views,* 98th Cong., 1st Sess., 2 December 1983, [hereinafter *FDA's Regulation of Zomax: Report*] 5.

8. Ibid., 7.

9. *FDA's Regulation of Zomax: Hearings,* 341.

10. *FDA's Regulation of Zomax: Report,* 14.

11. "Tolmetin-Induced Anaphylactoid Reactions: A review of 25 Cases," reprinted in *FDA's Regulation of Zomax: Hearings* 1983, 266.

12. Dale and Lemanowicz v. Johnson & Johnson, (D.N.J.), Case No. 89-4057 (DRD), Answer to the First Amended Complaint, Demand for Jury Trial, 19, ¶ 138.

13. Ibid., ¶ 139.

14. Patrick H. Seay, Memorandum to R. Z. Gussin, 8 September 1984, 9–10.

15. Dale and Ed Lemanowicz, v. Johnson & Johnson, (D.N.J.), Case No. 89-4057 (DRD), 19, 33, and Exhibit 40.

16. Dale v. Johnson & Johnson, Answer to First Amended Complaint, 10.

17. Chris Nichols, Persona Representative of Estate of Twila Ann Nichols v. McNeilab, Transcript of Depositions, 5 February 1990, 24–26.

18. J. W. Gorder, MD, Memorandum 7 July 1981, 3.

19. *FDA's Regulation of Zomax: Hearings,* 26–27 April 1983, 119.

20. Seay, Memorandum, 4.

21. *FDA's Regulation of Zomax: Hearings,* 300.

22. Ibid., 327.

23. Ibid., 462.

24. Dale v. Johnson & Johnson, Case No. 89-4057 (DRD), 31–32 and Exhibit 6.

25. Dale v. Johnson & Johnson, Answer to First Amended Complaint, 15, ¶ 104.

26. Dale and Lemanowicz v. Johnson & Johnson, Opinion, 18 May 1990, 3.

27. *FDA's Regulation of Zomax: Hearings,* 19.

28. "Estimated Incidence of Anaphylactic Reactions Reported for the Non-steroidal Anti-Inflammatory Drugs in the May 26, 1981 ADR Highlights," *FDA's Regulation of Zomax: Hearings,* 26–27 April 1983, 282.

29. Estate of Nichols v. McNeilab, Transcript of Depositions, 5 February 1990, 52.

30. *FDA's Regulation of Zomax: Hearings,* 12.

31. *FDA's Regulation of Zomax: Report,* 2.

32. Estate of Nichols, v. McNeilab, Transcript of Depositions, 5 February 1990, 92.

33. *FDA's Regulation of Zomax: Hearings,* 110.

34. Ibid., 109.

35. Dale v. Johnson & Johnson, Case No. 89-4057 (DRD), 50, ¶¶ 193 and 194.

36. Ibid., 48.

37. Ibid., 48, 50.

38. Ibid., 48, 50, and Exhibit 9: 2 and 12.

39. Dale v. Johnson & Johnson, Answer to First Amended Complaint, 48.

40. Dale v. Johnson & Johnson, Case No. 89-4057 (DRD), 61, ¶ 247, and Exhibit 14.

41. Dale v. Johnson & Johnson, Answer to First Amended Complaint, 32, ¶ 247.

42. *FDA's Regulation of Zomax: Report,* 3.

43. *FDA's Regulation of Zomax: Hearings,* 10–11.

44. *FDA's Regulation of Zomax: Report,* 3.

45. *FDA's Regulation of Zomax: Hearings,* 4.

46. Ibid.

47. *FDA's Regulation of Zomax: Report,* 3.

48. Ibid., 4–5.

49. Davis, Statement, 2.

50. Dale v. Johnson & Johnson, Answer to First Amended Complaint, 32, ¶ 251.

51. *FDA's Regulation of Zomax: Hearings,* 175.

52. *FDA's Regulation of Zomax: Report,* 4.

53. Manfred Ohrenstein, letter to Ted Weiss, 26 October 1983, 5.

54. Strategic Marketing Corporation, "Summary Report, Physician's Attitudes Towards Possible Reintroduction of Zomax, conducted for McNeil Pharmaceutical," February 1985, 5–7.

55. *FDA's Regulation of Zomax: Report,* 14.

56. *FDA's Regulation of Zomax: Hearings,* 20–21.

57. Ibid., 264.

58. *FDA's Regulation of the New Drug Versed: Hearings before a Subcommittee of the House Committee on Government Operations,* 100th Cong., 2nd Sess., 5, 10 May 1988, 15–16.

59. *FDA and the Medical Device Industry: Hearing before the Subcommittee on Oversight and Investigation of the H.R. Committee on Energy and Commerce,* 101st Cong., 2nd Sess., 26 February 1990, Serial No. 101-127, 196.

60. C. K. Mok, letter to John Dingell, 7 August 1989, in *The Bjork-Shiley Heart Valve: "Earn As You Learn,"* Staff Report prepared for the use of the Subcommittee on Oversight and Investigations of the House Committee on Energy and Commerce (101st Cong., 2nd Sess., February 1990), 106–107.

61. *FDA's Regulation of the New Drug Versed,* 34.

62. *FDA's Regulation of Zomax: Hearings,* 19.

63. Ibid., 425.

64. Ibid., 40.

65. *FDA and the Medical Device Industry,* 285.

66. Ibid., 497.

67. *The Bjork-Shiley Heart Valve: "Earn As You Learn,"* Staff Report prepared for the use of the Subcommittee on Oversight and Investigations of the House Committee on Energy and Commerce (101st Cong., 2nd Sess., February 1990), 17–18.

68. Food and Drug Administration, *Report of the Shiley Task Group,* December 1990, 6–7.

69. *FDA's Regulation of Zomax: Report,* 9, including footnote 10.

70. Ibid., 9–10.

71. Ibid., 10.

72. Ibid.

73. *FDA and the Medical Device Industry,* 504.

74. *Bjork-Shiley Heart Valve,* 6–7.

75. *FDA and the Medical Device Industry,* 281.

76. *The Wall Street Journal,* 7 November 1991, 1.

77. *Bjork-Shiley Heart Valve,* 113–114.

78. Ibid., 112.

79. Ibid., 22.

80. Ibid., 23.

81. FDA, *Report,* 12, 15–16.

82. *FDA and the Medical Device Industry,* 53.

83. *Bjork-Shiley Heart Valve,* 24.

84. FDA, *Report,* 17–18.

85. James A. Dale and Devra Lee Davis, "Court ordered secrecy and public health," *Lancet,* 21 July 1990.

86. Ibid.

87. J. A. Murray, Memorandum 17 March 1983.

88. Dale v. Johnson & Johnson, Answer to First Amended Complaint, 78, ¶ 319.

89. *Bjork-Shiley Heart Valve,* 1.

90. Ibid., 2.

91. *FDA and the Medical Device Industry,* 154–155.

92. *Bjork-Shiley Heart Valve,* 15.

93. "FDA Says Pfizer Inadequately Warned Heart-Valve Recipients of Risk of Death," *The Wall Street Journal,* 10 December 1990, B4.

94. *FDA and the Medical Device Industry,* 500.

95. Ibid., 501.

96. *FDA's Regulation of the New Drug Versed,* 71.

97. Ibid., 341.

98. Ibid., 73.

99. Ibid., 5–6.

100. Ibid., 11.

101. Ibid., 141.

102. Ibid.

103. Ibid., 238.

104. Ibid.

105. Ibid., 26.

106. *FDA's Deficient Regulation,* 32.

107. *FDA's Regulation of the New Drug Versed,* 100.

108. J. G. Reves, MD, Discussion of "Comparison of Midazolam and Diazepam for Sedation During Plastic Surgery," by Paul F. White et al., *Plastic and Reconstructive Surgery* (May 1988), reprinted in *FDA's Regulation of the New Drug Versed,* 255.

109. *FDA's Deficient Regulation,* 33.

110. *FDA's Regulation of the New Drug Versed,* 103.

111. Ibid., 141.

112. Ibid., 27, 257.

113. Ibid., 86.

114. Ibid., 560.

115. Ibid., 108, 561.

116. Ibid., 460.

117. Ibid., 478.

Chapter 6

1. Committee on Government Operations, "*Are Scientific Misconduct and Conflicts of Interest Hazardous to Our Health?*" 19th *Report together with Dissenting and Additional Views, September 10, 1990,* (Washington, DC: U.S. Government Printing Office, 1990) [hereinafter *Are Scientific Misconduct and Conflicts of Interest Hazardous?*], 6.

2. *Is Science For Sale? Conflicts of Interest vs. the Public Interest: Hearing before a Subcommittee of the House Committee on Government Operations,* 101st Cong., 1st Sess., 13 June 1989, 131.

3. *Are Scientific Misconduct and Conflicts of Interest Hazardous?* 44.

4. Ibid.

5. Ibid.

6. Subcommittee on Human Resources and Intergovernmental Relations, House Committee on Government Operations, Testimony of Lawrence M. Solomon, 11 June 1991.

7. Subcommittee on Human Resources and Intergovernmental Relations, House Committee on Government Operations, Statement of David A. Kessler, MD, FDA Commissioner, 11 June 1991, 2.

8. *Federal Response to Misconduct in Science: Are Conflicts of Interest Hazardous to Our Health," Hearing before a Subcommittee of the House Committee on Government Operations,* 100th Cong., 2nd Sess., 29 September 1988, 76.

9. Ibid., 75.

10. TIMI Study Group, "The Thrombolysis in Myocardial Infarction (TIMI) Trial," *New England Journal of Medicine,* 312: 932–936.

11. *Federal Response to Misconduct, Hearing,* 78–79.

12. *Are Scientific Misconduct and Conflicts of Interest Hazardous?* 22.

13. Ibid., 26.

14. *Federal Response to Misconduct,* 80.

15. *Are Scientific Misconduct and Conflicts of Interest Hazardous?* 27.

16. Ibid., 23.

17. Erdem I. Cantekin, Timothy W. McGuire, and Robert L. Potter, "Biomedical Information, Peer Review, and Conflict of Interest as They Influence Public Health," *Journal of the American Medical Association* 263 (9 March 1990): 1428.

18. *Federal Response to Misconduct,* 6.

19. Ibid.

20. *Is Science for Sale?* 273.

21. Ibid., 81.

22. *Federal Response to Misconduct,* 98.

23. *Is Science for Sale?* 269.

24. *Are Scientific Misconduct and Conflicts of Interest Hazardous?* 36–37.

25. *Federal Response to Misconduct,* 12.

26. Ibid., 46.

27. *Is Science for Sale?* 213.

28. *Are Scientific Misconduct and Conflicts of Interest Hazardous?* 36.

29. *Federal Response to Misconduct,* 21.

30. *Is Science for Sale?* 267–268 (reprint of Memorandum from Howard Hyatt, Director, Division of Management Survey and Review, OA to Deputy Director, NIH, 1 June 1989).

31. Federal Response to Misconduct, 7.

32. "Reply to the Medsger Report," 29 December 1987, 1.

33. *Federal Response to Misconduct,* 244 (reprint of letter to Earleen Elkins, NIH, 23 June 1987).

34. *Are Scientific Misconduct and Conflicts of Interest Hazardous?* 34–35.

35. *Federal Response to Misconduct,* 15.

36. Appeal letter to Thomas Detre, MD, 24 May 1989, 8.

37. Cantekin, McGuire, and Potter, "Biomedical Information, Peer Review, and Conflict of Interest," 1430, endnote 5.

38. *Federal Response to Misconduct,* 198.

39. *Federal Response to Misconduct,* 248–249 (reprint of letter to Earleen Elkins, 23 June 1987).

40. *Is Science for Sale?* 214.

41. Cantekin, McGuire, and Potter, "Biomedical Information, Peer Review, and Conflict of Interest," 1430, endnote 5.

42. Office of Scientific Integrity, "Inquiry covering clinical trials for otitis media conducted at the Children's Hospital of Pittsburgh," 25.

43. Erdem I. Cantekin letter to William Raub, PhD, Deputy Director, NIH, 27 June 1989, 2–4.

44. Erdem I. Cantekin, Comments and Criticism, attached to 13 August 1990 letter to Suzanne Hadley, 3.

45. Ibid., 9.

46. Ibid., 29.

47. OSI, "Inquiry," 29.

48. Jack Froom et al., "Diagnosis and Antibiotic Treatment of Acute Otitis Media: Report from International Primary Care Network," *British Medical Journal* 300 (1990): 582–586.

49. "Is Science for Sale?" 219.

50. Cantekin, Comments, 8.

51. *Federal Response to Misconduct,* 21.

52. Ibid., 13.

53. "Report of the Executive Hearing Board of the School of Medicine, Formal Inquiry into Allegations of Scientific Misconduct by Erdem I. Cantekin, PhD," sent to Dean George M. Bernier, Jr., 6 April 1989, 13.

54. Robert L. Potter, Letter to Thomas Detre, MD, 24 May 1989, 6.

55. Cantekin, McGuire, and Potter, "Biomedical Information, Peer Review, and Conflict of Interest," 1430, endnote 8.

56. "Report of the Executive Hearing Board," 13.

57. *Is Science for Sale,* 248–250 (reprint of letter from William Donaldson, MD, Medical Director, Children's Hospital, Edwin K. Zechman, President, Children's Hospital, and Eugene N. Myers, MD, Chairman, Department of Otolaryngology, University of Pittsburgh School of Medicine to Arnold S. Relman, MD, Editor, *The New England Journal of Medicine,* 1 August 1986, p. 2).

58. Cantekin, McGuire, and Potter, "Biomedical Information, Peer Review, and Conflict of Interest," 1429.

59. Arnold S. Relman, Letter to Robert Potter, 6 January 1987, 2.

60. *Is Science for Sale?* 200.

61. Ibid., 198.

62. Cantekin, Comments, 25.

63. "Report of the Executive Hearing Board," 10.

64. Robert L. Potter, Letter to Thomas Detre, MD, 6 April 1989, 1–2.

65. Watson committee, meeting of 16 February 1988, 5:15 P.M., 5 (verbatim transcript).

66. Letter, 21 March 1988, 5.

67. *Federal Response to Misconduct,* 200.

68. *Are Scientific Misconduct and Conflicts of Interest Hazardous?* 33.

69. Ibid., 32.

70. "Report of the Executive Hearing Board," 13.

71. "Report to the President of the University of the Appeal Panel Reviewing the Proceedings Involving Erdem I. Cantekin, PhD," April 1990, 14.

72. Robert L. Potter, Letter to Wesley W. Posvar, 23 April 1990, 8.

73. Charles D. Bluestone, MD, Letter to Jerome C. Goldstein, MD, 10 August 1987, 1.

74. Robert L. Potter, Letter to Wesley W. Posvar, 23 April 1990, 5.

75. Dissenting Opinion of Max A. Lauffer to Report to the President of the University of the Appeal Panel Reviewing the Proceedings Involving Erdem I. Cantekin, PhD, April 1990, 16.

76. George M. Bernier, Jr., MD, Letter to Erdem Cantekin, 24 April 1990.

77. OSI, "Inquiry," 54.

78. Telephone interview with the NIH/NIDCD (National Institute of Deafness and Communicative Disorders), 17 October 1991.

Chapter 7

1. Memorandum for Lt. General Watts, AF/AC Thru: Mr. Fitzgerald, Subject: Pamphlet entitled "What is IR&D/BP" published by the Electronic Industries Association, 18 August 1987, 4.

2. Department of Defense Inspector General, *Semiannual Report to the Congress,* April 1 1989 to September 30, 1989, pp. 3-11 to 3-14.

3. *The New York Times,* 27 July 1990, 1.

4. Cited in Joan Dopico Winston, "Defense-Related Independent Research and Development in Industry," Congressional Research Service, 18 October 1985, 35.

5. U.S. Congress, Office of Technology Assessment, *The Defense Technology Base: Introduction and Overview—A Special Report,* OTA-ISC-374 (Washington, DC: U.S. Government Printing Office, March 1988), 48.

6. Testimony of Bruce Chafin to the *Joint Hearing before the Subcommittee on Oversight and Investigations of the Committee on Energy and Commerce and the Subcommittee on Criminal Justice, House Committee on the Judiciary,* 100th Cong., 1st Sess., 30 July 1987, 14–15.

7. *Hearings before the Senate Committee on Armed Services,* 100th Cong., 2nd Sess., 13 April 1988, 392.

8. Memorandum for General Watts, SAF/AC, 16 May 1988, 3.

9. Defense Contract Audit Agency, "Summary, Independent Research and Development and Bid and Proposal Cost Incurred by Major Defense Contractors in the Years 1988 and 1989," March 1990, 1. The report defines a major contractor as one with auditable costs greater than $40 million during the fiscal year 1989.

10. OTA Report, 1988, 48.

11. Memorandum for Lt. General Watts, AF/AC, Thru: Mr. Fitzgerald, 12.

12. *Hearings before the Senate Committee on Armed Services,* 100th Cong., 2nd Sess., 18 March 1988, 78.

13. Quoted in Joan Dopico Winston, "Defense-Related Independent Research and Development in Industry," Congressional Research Service, 18 October 1985, 40.

14. Testimony of Dr. Michael Rich, *Hearings before the Senate Committee on Armed Services,* 100th Cong., 2nd Sess., 18 March 1988, 78.

15. Memorandum for Mr. Fitzgerald, Subject: IR&D and RAND Briefing, 16 May 1988, 1.

16. Memorandum for General Watts, 24 June 1988, 3.

17. *Hearings before the Senate Committee on Armed Forces,* 88.

18. U.S. General Accounting Office, "*Competition—Issues on Establishing and using Federally Funded Research and Development Centers,*" GAO/NSIAD-88-22, March 1988, 39.

19. Memorandum for Lt. General Watts, AF/AC Thru: Mr. Fitzgerald, 3.

20. *Hearing on the Harassment of A. E. Fitzgerald,* Subcommittee on Oversight and Investigations of the House Committee on Energy and Commerce, 6 November 1985 (transcript), 5–6.

21. Ibid., 210.

22. A. Ernest Fitzgerald, *The Pentagonists,* Houghton Mifflin, Boston, 1989.

23. *DOD Reorganization Implementation: Hearings before the Investigations Subcommittee of the House Committee on Armed Services,* 100th Cong., 1st Sess., 4 June 1987, 243.

24. Office of Assistant Secretary of Defense (Public Affairs), Press Release No. 544-89, 30 November 1989.

25. *Hearing before a Subcommittee of the H.R. Committee on Government Operations,* 98th Cong., 1st Sess., 22 September 1983, 4.

26. House Committee on Government Operations, "Additional views of Hon. Frank Horton, Hon. John N. Erlenborn, Hon. Lyle Williams, Hon. William F. Clinger, Jr., Hon. Raymond J. McGrath, Hon. Judd Gregg, Hon. Tom Lewis, Hon. Alfred A. (Al) McCandless, and Hon. Dan Schaefer to "Defense Science Boards: A Question of Integrity": *27th Report together with Additional Views,* 28 November 1983, 22–23.

27. U.S. General Accounting Office, "DOD Instruction 5000.5x, Standard Instruction Set Architecture for Embedded Computers," (MASAD-82-14) 27 January 1982.

28. "Additional Views": *Report,* Appendix III.

29. Ibid., 7.

30. *Hearing of the* House *Committee on Government Operations,* 77–78.

31. Ibid., 112.

32. Ibid., 27.

33. Robert J. Johnston, quoted in *House Report,* 28 November 1983, 6.

34. *Hearing of the* House *Committee on Government Operations,* 40.

35. Ibid., 68.

36. Ibid., 59.

37. *House Report,* 28 November 1983, 5.

38. Ibid., 10.

39. *SDI Program: Hearings before the Defense Policy Panel and Research and Development Subcommittee of the House Committee on Armed Services,* 100th Cong., 1st Sess., 8 July 1987, 194 (read by Congressman John M. Spratt, Jr.).

40. Memorandum for Under Secretary of Defense (Acquisition), Subject: Letter Report of the Defense Science Board Task Force Subgroup Strategic Air Defense—Strategic Defense Milestone (SDM) Panel, undated. Also discussed in "Pseudo-Science and SDI," Joseph Romm (with major assistance from Joe Cirincione of the staff of the House Committee on Armed Forces), *Arms Control Today,* October 1989, 18.

41. *The Washington Post,* 9 July 1987.

42. *Washington Times,* 9 July 1987, 4.

43. *SDI Program: Hearings before the Defense Policy Panel and Research and Development Subcommittee of the House Committee on Armed Services,* 100th Cong., 1st Sess., 8 July 1987, 197.

44. Rick Chervenak, on the Oversight and Investigation Subcommittee of the House Committee on Energy and Commerce, telephone interview with author, April 1990.

45. Report on the Activity of the House Committee on Energy and Commerce, 101st Cong., 1st Sess., 5 February 1990, 213.

46. Memorandum for the Secretary of the Navy Via: Assistant Secretary of the Navy (Research, Development and Acquisition), Subject: A-12 Administrative Inquiry, 28 November 1990, 1.

47. Ibid., 1.

48. Ibid., 19.

49. Ibid., 5.

50. Ibid., 9.

51. Ibid., 7.

52. Ibid., 8.

53. United States of America v. Ronald Emile Brousseau, No. Cr 85-387-JMI, Sentencing Memorandum.

54. "Report on Guidelines for Determining the Degree of Risk Appropriate for the Development of Major Defense Acquisition Systems, and Assessing the Degree of Risk Associated with Various Degrees of Concurrency; and Concurrency in Major Acquisition Programs," 10 April 1990.

55. Ibid., 1–2.

56. U.S. General Accounting Office, "Weapons Testing—DOD Needs to Plan and Conduct More Timely Operational Tests and Evaluation," GAO/NSIAD-90-107, May 1990, 5.

57. U.S. General Accounting Office, "*Department of Defense: Improving Management to Meet the Challenges of the 1990s,*" GAO/T-NSIAD-90-57, 13.

58. Vice Admiral Robert R. Monroe, USN (RET.) "The Issue of 'Concurrency' in Weapons System Acquisition," April 1984 (privately circulated paper).

59. Ibid., 4.

60. U.S. General Accounting Office, "Production of Some Major Weapon Systems Began With Only Limited Operational Test and Evaluation Results," GAO/NSIAD-85-68, 19 June 1985, 4.

61. Ibid., 20.

62. GAO, "Weapons Testing," 13.

63. Ibid., 13.

64. U.S. General Accounting Office, "Weapon Systems: Concurrency in the Acquisition Process," GAO/T-NSIAD-90-43, 7.

65. Ibid., 3–4.

66. Monroe, "Concurrency," 2.

67. General Accounting Office, "Apache Helicopter—Serious Logistical Support Problems Must Be Solved to Realize Combat Potential," GAO/NSIAD-90-294, 32.

68. "Improvements Needed in Development Testing," B-163058, 13.

69. Ibid., 1.

70. Subcommittee on Oversight and Investigations of the House Committee on Energy and Commerce and the Subcommittee on Criminal Justice, House Committee on the Judiciary, 100th Cong., 1st Sess., 28 October 1987, 114.

71. Ibid., 134–135.

72. U.S. General Accounting Office, "Weapons Testing—Quality of DOD Operational Testing and Reporting," GAO/PEMD-88-32BR, 19.

73. U.S. General Accounting Office, "Weapon Performance—Operational Test and Evaluation Can Contribute More to Decisionmaking," GAO/NSIAD-87-57, December 1986, 4.

Chapter 8

1. *Hearing on Indirect Charges at Stanford University,* Subcommittee on Oversight and Investigations of the House Committee on Energy and Commerce, transcript, 13 March 1991, 36.

2. Ibid., 9.

3. Ibid., 141–142.

4. Ibid., 84–85.

5. Ibid., 105.

6. Ibid., 111–112.

7. Ibid., 145.

8. Ibid., 37.

9. Ibid., 36.

10. Arnold S. Relman, "Economic Incentives in Clinical Investigation," *The New England Journal of Medicine* 320 (6 April 1989): 933–934.

11. Bernadine Healy, MD, et al., "Special Report—Conflict-of-Interest Guidelines for a Multicenter Clinical Trial of Treatment after Coronary–Artery Bypass-Graft Surgery" 320 *The New England Journal of Medicine,* (6 April 1989): 949–951.

12. Relman, "Economic Incentives," 933–934.

13. John Crewdson, faxed letter to Storm Whaley, 6 July 1990, (p. 2, Document B-1 of *Findings and Required Actions Regarding Allegations of Noncompliance with HHS Regulations for the Protection of Human Research Subjects Involving the National Institutes of Health Intramural Research Program,* Office for Protection from Research Risks, Division of Human Subject Protection, 31 May 1991).

14. Office for Protection from Research Risks, Division of Human Subject Protection, *Findings and Required Actions Regarding Allegations of Noncompliance with HHS Regulations for the Protection of Human Research Subjects Involving the National Institutes of Health Intramural Research Program,* 31 May 1991, 2.

15. "After 3 Die in Tests, France Bans AIDS Vaccine," *Chicago Tribune,* 16 June 1991, 1, 10.

16. Office for Protection, *Findings and Required Actions,* 9.

17. Ibid., 13.

18. Ibid., 20.

19. "Error, Fraud and Misconduct in Science," The Arthur Miller Lecture on Science and Ethics, The Context Support Office, The Program in Science, Technology, and Society, MIT, April 7, 1990.

20. *Whistleblower Protection: Hearings before the Subcommittee on Civil Service of the House Committee on Post Office and Civil Service,* 99th Cong., 1st Sess., 15 May, 18, 26 June, 1985 (Serial No. 99-19), 25–28.

21. *Hearings on the Role of General Dynamics and Westinghouse in Selling Support Equipment to the Air Force for the F-16 Aircraft,* Subcommittee on Oversight and Investigations, House Committee on Energy and Commerce, 23 September 1985 (transcript), 67–68.

22. *Whistleblower Protection,* 263.

23. *Hearing on Apparent Financial Wrongdoing by an Official in the Laboratory of Tumor Cell Biology at the National Institutes of Health,* Subcommittee on Oversight and Investigations, House Committee on Energy and Commerce, 6 March 1991 (transcript), 2–9.

24. *DOD Reorganization Implementation: Hearings before the Investigations Subcommittee of the H.R. Committee on Armed Services,* 100th Cong., 1st Sess., 28 April, 1, 28 May, 3, 4, 18, 23 June, and 4 November 1987, 240.

25. *Hearing on Recent Action Involving the Office of Scientific Integrity of the National Institutes of Health and Events Regarding the Cleveland Clinic,* Subcommittee on Oversight and Investigations, Committee on Energy and Commerce, August 1, 1991, transcript, pp. 5–6.

26. H.R. Report No. 1819, 102nd Cong., 1st Sess.

27. Quoted from "Courage Without Martyrdom," material supplied by the Government Accountability Project, Washington, DC.

Index